Hidden
Natural
Histories

TREES

Noel Kingsbury is a best-selling horticulturalist and writer. He is the author of many books, including *Designing with Plants, Natural Gardening in Small Spaces, Hybrid: The History and Science of Plant Breeding,* and *Gardening with Perennials: Lessons from Chicago's Lurie Garden*—the latter two also published by the University of Chicago Press—as well as the coeditor of *Vista: The Culture and Politics of Gardens.* He lives and gardens in western England near the world-famous book town of Hay-on-Wye.

First published in the United States of America in 2015 by
The University of Chicago Press Chicago 60637

Copyright © 2015 Quintessence Editions Ltd.

24 23 22 21 20 19 18 17 16 15   1 2 3 4 5

ISBN-13: 978-0-226-28221-3   (paper)
ISBN-13: 978-0-226-21589-1   (e-book)
DOI: 10.7208/chicago/9780226215891.001.0001

Library of Congress Cataloging-in-Publication Data

Kingsbury, Noel, author.
  Hidden natural histories. Trees / Noel Kingsbury.
     pages cm
  ISBN 978-0-226-28221-3 (paperback. : alkaline paper) — ISBN 0-226-28221-X (paperback. :
alkaline paper) — ISBN 978-0-226-21589-1 (e-book) — ISBN 0-226-21589-X (e-book) 1.
Trees—Utilization. I. Title.
  QK475.K54 2015
  582.16—dc23
                                                                    2014037013

This book was designed and produced by
Quintessence Editions Ltd. The Old Brewery, 6 Blundell Street, London N7 9BH

Project Editor	Zoë Smith
Designer	Isabel Eeles
Editors	Frank Ritter, Fiona Plowman
Production Manager	Anna Pauletti
Editorial Director	Jane Laing
Publisher	Mark Fletcher

Color reproduction by Colourscan Print Co Pte Ltd, Singapore
Printed and bound in China by Shanghai Offset Printing Products Ltd.

 ∞ This paper meets the requirements of ANSI/NISO Z39.48-1992 (Permanence of Paper).

Hidden
Natural
Histories

NOEL
KINGSBURY

TREES

THE SECRET PROPERTIES OF **150 SPECIES**

THE UNIVERSITY OF CHICAGO PRESS

CHICAGO

Trees by Common Name

Contents

How to Use This Book

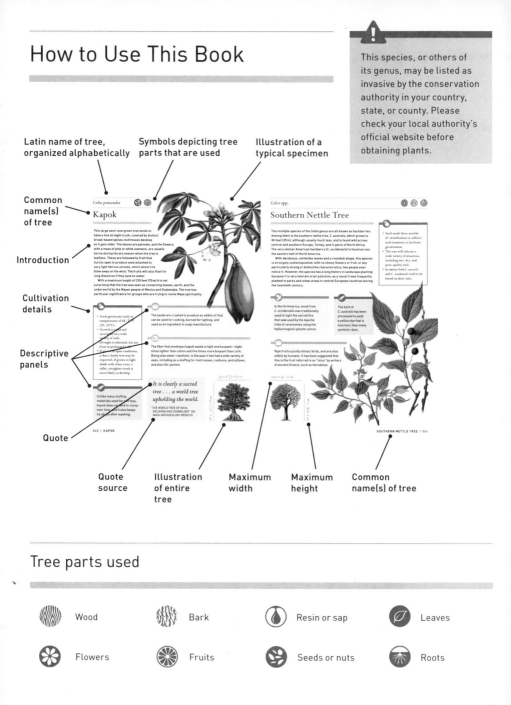

This species, or others of its genus, may be listed as invasive by the conservation authority in your country, state, or county. Please check your local authority's official website before obtaining plants.

Latin name of tree, organized alphabetically

Symbols depicting tree parts that are used

Illustration of a typical specimen

Common name(s) of tree

Introduction

Cultivation details

Descriptive panels

Quote

Quote source

Illustration of entire tree

Maximum width

Maximum height

Common name(s) of tree

Ceiba pentandra

Kapok

This large semi-evergreen tree tends to have a fine straight trunk, covered by distinct broad-based spines; buttresses develop as it gets older. The leaves are palmate, and the flowers, with a mass of pink or white stamens, are usually borne during the dry season when the tree is leafless. These are followed by fruit that bursts open to produce seed attached to very light fibrous strands, which allow it to blow away on the wind. The fruits will also float for long distances if they land on water.
With a maximum height of 230 feet (70 m) it is not surprising that the tree was seen as connecting heaven, earth, and the underworld by the Mayan people of Mexico and Guatemala. The tree has particular significance for groups who are trying to revive Maya spirituality.

• Seeds germinate easily at temperatures of 68–77°F (20–25°C).
• Growth is rapid and successful on a wide variety of soils.
• Drought is tolerated, but not frost or prolonged cold.
• If grown in open conditions, a short, bushy tree may be expected; if grown in light shade with other trees, a taller, straighter trunk is more likely to develop.

The seeds are crushed to produce an edible oil that can be used for cooking, burned for lighting, and used as an ingredient in soap manufacture.

The fiber that envelopes kapok seeds is light and buoyant—eight times lighter than cotton and five times more buoyant than cork. Being also water-repellent, in the past it has had a wide variety of uses, including as a stuffing for mattresses, cushions, and pillows, and also life-jackets.

Unlike many stuffing materials used for soft toys, kapok does not tend to clump over time, and it also keeps its shape after washing.

It is clearly a sacred tree . . . a world tree upholding the world.

"THE WORLD TREE OF MAYA RELIGION AND COSMOLOGY" ON MAYA ARCHAEOLOGY WEBSITE

060 | KAPOK

Celtis spp.

Southern Nettle Tree

The multiple species of the *Celtis* genus are all known as hackberries. Among them is the southern nettle tree, *C. australis*, which grows to 80 feet (25 m), although usually much less, and is found wild across central and southern Europe, Turkey, and in parts of North Africa. The very similar American hackberry (*C. occidentalis*) is found across the eastern half of North America.
With deciduous, nettlelike leaves and a roundish shape, this species is strongly undistinguished, with no showy flowers or fruit, or any particularly strong or distinctive characteristics; few people even notice it. However, the species has a long history in landscape planting because it is very tolerant of air pollution, as a result it was frequently planted in parks and urban areas in central European countries during the twentieth century.

• Seed needs three months of stratification or sulfuric acid treatment to facilitate germination.
• The tree will tolerate a wide variety of situations, including wet, dry, and poor-quality soils.
• In nature both *C. australis* and *C. occidentalis* tend to be found on drier soils.

In North America, wood from *C. occidentalis* was traditionally used to light the sacred fire that was used by the Apache tribe in ceremonies using the hallucinogenic peyote cactus.

The bark of *C. australis* has been processed to yield a yellow dye that is less toxic than many synthetic dyes.

Ripe fruits quickly attract birds, and are also edible by humans. It has been suggested that this is the fruit referred to as "lotus" by writers of ancient Greece, such as Herodotus.

SOUTHERN NETTLE TREE | 061

Tree parts used

Symbol	Part
	Wood
	Bark
	Resin or sap
	Leaves
	Flowers
	Fruits
	Seeds or nuts
	Roots

Key to descriptive panels

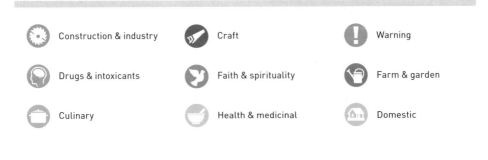

Construction & industry

Craft

Warning

Drugs & intoxicants

Faith & spirituality

Farm & garden

Culinary

Health & medicinal

Domestic

Systematic, large-scale exploitation of certain trees: for timber by the building industry, and, particularly after the Industrial Revolution, for resins and other wood-based products.

In the distant past, crafts people fashioned selected raw materials from local trees into everything their communities needed, from roofing and fencing to baskets and wooden bowls.

Certain trees present particular risks to humans. Many contain toxins, evolved as defences against insects or grazing animals. Others may become invasive and threaten local ecologies.

Plant chemicals can be psychoactive—they change the way the human brain works. The human race has long sought out certain tree products, such as cocoa, for the pleasures they offer.

Trees have always played a part in human spiritual life. Some were seen as embodying spirits or gods. Others produced substances, such as frankincense, that had spiritual associations.

Trees have long been cultivated around the home, by gardeners, rural families, and smallholder farmers, for their specific benefits: food, shade, wind protection, animal fodder, and others.

From the nuts and acorns collected by the earliest peoples to exotic fruits imported from distant lands, tree-borne foods have been of inestimable value in sustaining humankind.

Traditional medicine relies on particular trees for some of its active ingredients, which may be extracted in varying potencies from the bark, inner tissue, roots, leaves, flowers, or fruits.

Tree products are found everywhere in domestic life, from firewood and substances that repel clothes moths to rare, precious woods used to make musical instruments.

Introduction   *Noel Kingsbury*

This book is about trees, and primarily about what makes trees useful to us. When we think of trees, and the products we get from them, our first thought is usually of wood. There is so much else, however. This book is aimed at helping the reader to understand the incredible diversity of our uses of trees, how dependent we have been on them throughout history, and how much our future might be tied up with them.

Much of our use of trees is hidden, in the sense that it is taken for granted, or not realized, hence the idea of the "hidden history." In many cases there is a continuity between the past and the present—for example, we still make great use of oak species for furniture, just as we have done more or less since furniture was invented in the dim and distant past. Many other uses have fallen by the wayside, and it is likely that most of them will never be revived. Other uses will be rediscovered, and almost certainly completely new ones will be chanced upon, too. It is the intention of this book to explore the past, present, and possible future uses of trees and make connections across time. First, we'll look at what defines a tree, and then explore a narrative about how trees grow in nature; we will draw parallels between the tree's natural place in the world and the story of humanity's involvement with trees.

What is a tree?

A tree may be defined as a woody plant with a single, clearly defined trunk. Both conifers and broad-leaved trees, although only very distantly related, share a similar habit of growth—they have an outer living layer, just below the bark, the cambium, which constantly lays down cells containing compounds that form the hard, strong structures known to us as wood. This younger wood is known as sapwood; although functionally dead, it conducts water up from the roots to the branches and leaves. Eventually the sapwood ages, stops conducting water, and hardens to become heartwood, the main function of which is supportive. It helps to think of a tree as a living skin of tissue spread over a hard, dead structure. Apart from the leaves, flowers, and fruit, this living tissue is sheltered beneath another layer of dead material: the bark.

Other plants that we call trees organize themselves very differently. The trunks of palms do not get wider as the plant ages, and all new growth is at the top. There are a few other oddities, too, such as the extraordinary looking Ombú trees of South America (*Phytolacca dioica*), which have a different tissue structure, and, included here, giant cacti that form internal woody supportive structures and, to all intents and purposes, have a treelike habit of growth, such as the saguaro (*Carnegiea gigantea*) of the southwestern United States.

Shrubs differ from trees in having multiple stems, often produced continually from the base. There is no clear dividing line between trees and shrubs, and included in this book are several species that are very much on the borderline between the two.

Pioneers and climax species

Land that is clear of trees, and has a soil and a climate that can support tree growth, does not stay clear of trees for very long. Unless regularly mown, grazed, or burned, seedlings of trees will soon appear in such places, even toxic and polluted postindustrial environments. Given time they will multiply, and a forest can appear remarkably quickly. The species that dominate at this stage tend to be fast-growing, and, quite often, relatively short-lived—these are the pioneer species. After a few decades they begin to be replaced by other species, generally slower-growing ones. The composition of the young forest can change markedly as the pioneers begin to be replaced, either by new arrivals or by slower-growing species that started at the same time as the pioneers but that are better adapted for the long haul. This process, whereby one range of species replaces another in a relatively predictable pattern, is known to ecologists as succession.

As the forest matures, pioneers are replaced by climax species. These are the ones that will dominate an environment in a particular climate zone for long periods of time and form a relatively stable plant community. Many climax species are very long-lived, and, because their seedlings are tolerant of shade, young ones can start to grow in the shadow of their parents. Thus, the stability of mature forest is only an illusion. Change is constant, if slow. Trees fall, or are felled, in which case the sudden arrival of light on the forest floor stimulates seedlings to rush for the light. These gaps provide the only chances in mature climax forest for pioneer species to establish themselves again, because their seedlings almost invariably need good light and have little tolerance of shade. Climax forest in much of temperate Eurasia and North America has had a long history of human intervention, with felling being followed by rapid regeneration or replanting. In other parts of the world, much of the forest is old-growth or virgin forest that has never undergone felling, and where regeneration is natural and slow.

Humanity and the global mix of tree species

In the past we tended to use trees of the mature climax community for our timber and other needs. Now we are at a point where there is not much of this left, and we are turning to the pioneer species for our construction and industrial requirements. The global species mix is changing as a result.

People have always used trees—this is what this book is about—but they have also seen trees in a negative way, as something in the way. Ever since humanity began to depend on arable crops like rice, corn, and wheat, or indeed pastoral farming based on grass-eating animals like sheep or cattle, forest has frequently been seen as expendable. The world's forests have endured huge levels of destruction as a result of clearing for agriculture, as well as unsustainable forest management, particularly cutting trees for firewood. Today the focus is on clearing forest, for oil palm plantations in Southeast Asia or ranching in South America. The past has seen vast areas of forest destroyed, often in the deep historical past—in South America by pre-Columbian peoples, for example. The Chinese and Indian classical civilizations came at the cost of huge areas of forest, while the Romans initiated a long period of forest clearing in Europe and around the Mediterranean. Yet today, forests are advancing in Europe, as they are in North America and in parts of Africa. Our relationship with trees and forests is clearly a dynamic one.

Exploitation and regeneration

In particular, old-growth forests new to exploitation have been squandered. For example, in the early days of settlement in North America, vast amounts of timber were used in huge open fires to warm log cabins built of extravagant amounts of wood. City streets were lined with wooden boardwalks; industrial furnaces were fired with trees. The forest seemed endless. More recently, extensive areas of rain forest in Myanmar (Borneo) have been felled to make formwork for reinforced concrete in Japan. The wood tends to be used only once, and the job could have been done just as effectively with plywood.

Today, in some regions more trees are growing and being planted than felled, and that is clearly a good thing. But which trees are being planted? In many cases the species may be very different to the ones that were there originally. When mature forest is felled, the trees cleared are the long-lived, often slow-growing, climax species. These are the species with the highest-quality timber—slow growth tends to produce wood that is relatively hard, dense, and strong. But when forest re-establishes itself naturally, it is the pioneer species, the fast-growing, sometimes almost weedy, species that predominate in the early stages. When forest is planted, also, it is often these fast-growing species that are selected, perhaps for a quick return, for firewood, or to reduce soil erosion. The result is a very different kind of forest community.

As old-growth timber begins to run out, the law of supply and demand pushes prices up. Tree conservation organizations campaign against forest destruction, and often countries possessing forest begin to value their resources more. All this means that the supply of high-quality timber from mature forest becomes restricted. In some cases, it becomes limited for the simple reason that a species has been driven into commercial extinction, as has happened with mahogany, brazilwood, and ipé. The timber trade and other users of forest products turn toward other lower-quality trees. Crucially, however, enterprising landowners and others start to plant. Some high-quality species can be grown relatively easily, and the trees are harvestable quickly enough to make planting them worthwhile. In tropical Asia, teak and shala (*Shorea robusta*) are two good examples of fast-growing quality timber trees, whereas mahogany, brazilwood, and ipé have so far proved to be uneconomic. Generally, though, fast-growing pioneer species are favored for planting.

As our exploitation of trees turns from the wholesale clearing of entire landscapes to a more sustainable "plant and harvest" approach, there is another development. High-quality timber from slower-growing trees gets expensive and is restricted to high-end uses. Other, cheaper timbers are selected for other purposes, and in some cases can even become fashionable. Mahogany furniture, for example, is now very unfashionable, while pine furniture remains popular.

Much of the lower-quality timber from relatively fast-growing trees will come from agroforestry. The basic concept is that new tree planting can be combined with other uses of land, such as grazing animals. Trees may be interspersed with crops that tolerate, or indeed need, some shade, like coffee or cardamom. A particularly important aspect of agroforestry is its production of forage for livestock.

Agroforestry offers a great many advantages to smallholders, and so is seen as offering potential to help lift rural people out of poverty. However, there is a risk of introducing aggressively spreading, potentially invasive, species to new areas. Successful agroforestry schemes help to protect surviving areas of old-growth forest from exploitation, but their very presence and short-term success may prevent the replanting, or the natural re-establishment, of native climax species of the area.

The future of old-growth forests

The human race is getting wealthier, and our concerns are shifting from survival to quality of life. Our respect for trees, and the diversity and magnificence of old-growth forest, is growing, although ironically at the same time we are demanding more resources, which often means more timber and other tree-derived products. The desire to conserve is very much a function of societies confident of their material well-being. We can only hope that a portion of the remainder of the world's old-growth forests survives into an era when enough people who admire them will concern themselves with protecting them. How we continue to exploit trees for their multitude of uses will play a major role in how the world's forests will look in centuries to come.

Abies balsamea

Balsam Fir

This evergreen conifer grows to 65 feet (20 m) and forms a large proportion of the boreal forest of eastern Canada and neighboring U.S. states. Its needles are typical of firs, being softer than those of the similar-looking spruces. The bark is smooth on young trees but sometimes marked by resin blisters; in older trees it tends to be fissured. The cones are upright on the branches, whereas spruce cones hang down. The seed in the cones is an important food source for squirrels and seed-eating birds, while the foliage is eaten by a number of moth species.

Balsam fir is one of several species that may be seen collectively sculpted into "fir waves" in mountainous areas. Consisting of bands of trees of different ages, the fir-wave phenomenon is a result of bitterly cold winds killing trees around gaps, with regeneration happening in the lee of the prevailing wind. Over time the wind effectively mows the trees into waves.

- Propagation is by seed, but viability is less than one year and germination percentages tend to be low.
- Germination occurs in spring; seedlings are notably shade tolerant.
- Adaptable to a wide range of soils, but appears to thrive best on damp soils.

In the nineteenth and twentieth centuries the resin was used as an adhesive. Applied dissolved in a solvent, the resin stays amorphous as it hardens, which means that it has excellent optical qualities. Since it became effectively invisible, it was widely used for gluing lenses and microscope slides. It has also been used as an invisible mend for chipped car windshields.

Up to 30 ft. (9 m)

50– 65 ft. (15–20 m)

Renowned for its fine shape, scent, and propensity to keep its needles, balsam fir is highly regarded for use as a Christmas tree. Even so, compared to spruce, its share of the market is relatively small.

Native American peoples are recorded as having made extensive use of balsam fir resin against worms, colic, urinary infections, and influenza. Until the late twentieth century, the resin was a popular ingredient in Western medicine for the treatment of colds and other respiratory problems.

Abies grandis

Grand Fir

A large conifer growing typically to 130–230 feet (40–70 m), the grand fir is native to western North America. It is a co-dominant species with Douglas fir in one part of its range, the Grand Fir/Douglas Fir ecoregion, occurring elsewhere as a minority element. Specimens found in Scotland are among the tallest of European trees. The needles have the typical flattened shape of *Abies* species and are arranged in an even plane.

Two distinct varieties are recognized: *Abies grandis* var. *grandis*, the coast grand fir, is the larger and faster-growing variety, and is valued by the timber trade, where it is often referred to as the "hem fir"; *Abies grandis* var. *idahoensis*, or interior grand fir, is smaller and has half the growth rate of its relative.

- Seed is best sown in fall, with the seedlings emerging in spring.
- Low germination percentages are to be expected of the seeds.
- The seedlings are relatively tolerant of shade and adult plants flourish on north-facing slopes, even at higher latitudes.
- Seedlings should be planted out when small, less than 35 inches (90 cm) high.
- A wide variety of conditions is tolerated, but the species prefers moist, well-drained, slightly acidic soils.

In North America, "coast grand fir" is noted both for being strong enough to use in construction and for its good appearance—very light, with an even grain; it is popular for furniture, framing, and moldings; if treated, it can be used for decking. In Europe it is regarded as a very soft wood, and is used for little more than a source of pulp.

The grand fir evolved a resin that deters moths from eating the foliage, and in the past people used the resin-scented foliage as a moth repellent. The wood can be burned as incense. A pink dye may be obtained from the bark, and the bark itself was used as a covering for canoes.

The inner bark is edible, and although it is not highly nutritious it can be ground up as flour and used for thickening soups or mixing with grains for baking bread. Otherwise, it tends to be used only as an emergency food, in the absence of anything else. The gum may be hardened with cold water and chewed.

Up to 20 ft. (6 m)

130–230 ft. (40–70 m)

Mimosa, Silver Wattle

Mimosa is a pioneer tree that quickly grows to 100 feet (30 m) but whose lifespan rarely exceeds thirty years. In nature mimosa is one of the species that rapidly establishes after fire in its homeland of southeast Australia. Elsewhere, it has become a rapidly spreading alien along roadsides and on disturbed land, and is generally seen as arboreal weed. However, in the south of France it has become a favorite tree, largely because of the fluffy, heavily fragrant yellow flowers borne at the end of the year.

- The hard seeds should be chipped with a sharp knife or boiled in water for a minute, in order to damage the coating sufficiently to allow water to be absorbed.
- Germination then occurs within weeks, and in any sunny situation the tree will grow rapidly. Pruning is not possible.
- Any soil is suitable for mimosa as long as it is not too wet.
- The species is hardy to 10°F (-12°C).

Like Australian blackwood (opposite), mimosa wood can be used in turning to make furniture and other items.

In the early twentieth century, the area around the French Riviera built up a considerable industry in exporting cut mimosa flowers to customers who appreciated them as a sign of the end of winter. Before dispatch, the flowers were kept for a day in a room heated to 72–77°F (22–25°C) at high humidity; this would considerably lengthen their lifespan.

Trees rapidly colonize waste ground, making it difficult for locally native species to re-establish themselves there. Places badly affected by the spread of mimosa include South Africa, India, and Madeira.

Up to 35 ft. (10 m)

70–100 ft. (21–30 m)

Acacia melanoxylon ⚠

Australian Blackwood

Renowned as a source of high-quality timber, this evergreen tree reaches 130 feet (40 m) and is found along the eastern seaboard of Australia. Like many Australian acacias it has phyllodes rather than leaves; these are technically stems that have taken the place of leaves. The flowers appear as ball-shaped clusters of pale yellow stamens.

The tree prefers moister habitats and generally the tallest in Australia are found in Tasmania. It has been introduced to many places with a similar climate where it is now regarded as a potentially invasive alien; it is particularly problematic in the Azores.

- Seeds require heat-shocking to induce germination. Immerse them in boiling water for a minute before sowing at 77°F (25°C).
- Moister soils are preferred; regions receiving less than 24 inches (60 cm) of rain a year are unsuitable for Australian blackwood.
- Frost tolerance varies with the provenance of the seed.
- The tree has moderate salt tolerance.

The tree's sap, either milky or reddish, is a source of gum acacia, or gum arabic.

THE UNIVERSITY OF SOUTHERN CALIFORNIA

The fine-grained wood is regarded as the Australian equivalent of walnut, suitable for quality furniture, inlays, and veneers. Naturally variable in color, often with contrasting bands, it also takes stains well. The wood can be steamed, which allows it to be shaped.

Up to 50 ft. (15 m)

Up to 130 ft. (40 m)

Given its good acoustic qualities and stability, Australian blackwood is used in the manufacture of musical instruments, including drums, guitars, violin bows, and even organ pipes. Its qualities fall between rosewood and mahogany.

The pollen-rich flowers are edible and can be used in a variety of dishes, such as fritters. The seeds can be ground and used as a substitute flour in cake and cookie recipes; they are being researched as a potential food in famines.

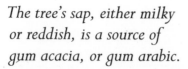

European Sycamore

This important deciduous woodland tree, with palmate leaves and very distinctive winged seed, has become naturalized well beyond its homeland in central and southern Europe. Wide-spreading and growing to a height of 115 feet (35 m), it is generally found as a component of mixed deciduous forests. In northern Europe it has shown itself to be an aggressive pioneer species, although concerns about its competitiveness with native species have lessened over time.

Sycamore is usually seen as a coarse tree, with an unattractive, heavy leaf drop and an ability to spread fast-growing seedlings over wide areas. The species is extremely windproof, however, and has long been favored for providing shelter for farms in exposed regions. The sycamore thrives in urban areas, too, where its prolific seedlings rapidly fill abandoned ground.

- Fresh seed germinates rapidly.
- Young trees have little tolerance of shade and should be grown in full sunlight.
- Given the size of the leaves and rapid growth, sycamore is unsuitable for hedges or any kind of clipping.
- Sycamore is susceptible to a range of fungal diseases that temporarily disfigure the foliage.

Sycamore wood is pale, fine-grained, and easy to work. Occasionally, trees develop a distinct pale, wavy pattern that runs at right angles to the grain; wood of this kind is known as "rippled" and is popular for veneers and the manufacture of fine furniture and musical instruments. Rippled sycamore also makes distinctive flooring.

Sycamore is a good wood for building kitchen furniture, or making butchers' blocks or wooden food containers, because it does not stain easily, and neither does it have any odor that could adversely affect food.

Up to 70 ft. (21 m)

Up to 115 ft. (35 m)

Sycamore's rapid growth and tolerance of exposure to cold and salt-laden winds makes it a very useful windbreak tree. It is a European tradition to plant it around dairy farms, where its open, spreading branches and heavy foliage ensure that cows are shaded as they wait for milking.

> *Although sycamore is not a native tree, it provides an important link in the food chain for many animals.*
>
> THE WILDLIFE TRUST'S WEBSITE

The earliest written accounts of maple sugaring were made in the early 1600s.

V. A. LUZADIS AND E. R. GOSSETT, "SUGAR MAPLE"

Acer saccharum

Sugar Maple

This large deciduous tree is part of the mature forest community of the northeastern United States and over the border into Canada. Growing to 115 feet (35 m), it casts a dense shade, tending to suppress the growth of any other plants beneath it; consequently, it is one of the few species of the region that can form almost pure stands, although more usually it occurs in conjunction with basswood, beech, and hickory species. A mature tree has a dense network of fibrous roots around it, but also deeper-thrusting roots that are able to access moisture lower down.

As with several maple species, sap rising from the roots can be tapped in spring by cutting into the tissue beneath the bark and attaching a bucket to collect the sap. The sugar maple's productivity and quality, however, make its maple syrup supreme, and the basis of an important industry in rural areas, with the Canadian state of Quebec supplying around three-quarters of all the maple syrup produced. In fall the foliage turns yellow, sometimes orange, contributing to the annual "autumn color" phenomenon for which the region is famous.

- Fresh seed (ideally picked straight from the tree and not dried) should be sown in fall to chill over winter, with germination occurring in the spring.
- Best in deep, fertile soils, but in practice it thrives on any moderately fertile soil.
- Young trees grow rapidly and are remarkably tolerant of being shaded, but they themselves cast a deep shade as they mature.

The syrup of the sugar maple is one of the world's most distinctive and delicious food products. The flavor is regarded as unique, with a complex chemistry that is not yet fully understood. Around 10 gallons (40 liters) of sap are needed to make ¼ gallon (1 liter) of syrup, through a process of boiling off water. Large quantities of the syrup are consumed neat, but the product is also processed into a wide range of "maple"-flavored products.

Known as "hardrock maple," the tree's timber is regarded as one of the best that North American forests have to offer, being hard, dense, and good-looking. It is ideal for flooring, and for making basketball courts, bowling alleys, baseball bats, pool cues, and skateboards, but is also flexible enough to make bows for archery, and bows for violins. Marked forms, such as the well-known "birds-eye" maple, are particularly sought after for furniture and flooring.

Being dense, sugar maple wood burns well and very hot. Used for smoking meats such as pork, the wood imparts flavor and sweetness.

The sap, served on its own, makes a refreshing drink—this is how it would have been consumed by Native Americans or pioneers. On nights when the temperature was likely to fall below freezing point they would partially concentrate the sap simply by leaving a bucket of it outside, and lifting out and discarding the ice that had formed on it by morning.

Up to 50 ft. (15 m)

Up to 115 ft. (35 m)

Baobab

Baobab is one of the most distinctive and characterful of trees, being immediately recognizable by its immense, swollen trunk. Although there are populations in northern India and the Arabian peninsula, it is largely an African species. It is widely distributed across the drier regions of that continent, where it has become a symbol of the African landscape. It is typically found as isolated specimens in savanna or relatively open woodland largely composed of other species. The height is usually no more than 80 feet (25 m) but, unusually, trees may develop a girth of similar proportions. The tree survives dry conditions because its trunk contains pithy tissue that stores water; the baobab is the world's largest succulent. The branches emerging from the top have the appearance of roots, particularly in the dry season when they are leafless, hence the popular name of "upside-down tree."

- Baobab is easy to grow from seed. Soak seeds in warm water for twenty-four hours before sowing in an open compost at a temperature of 68°F (20°C).
- Young baobab plants have little resemblance to mature ones because the leaves are simple (not compound) and the trunk is narrow, as in a normal tree.
- They may be grown in any well-drained soil, but the climate must be frost free. The tree is adapted to dry conditions and will not tolerate saturated ground.

The baobab is known as the "Mother of the Sahel" in sub-Saharan Africa for its nutritional and medicinal bounty. The fruit, called "monkey bread," has more than ten times the vitamin C content of oranges, while its pulp has been shown to have analgesic and anti-inflammatory qualities. The fruit can be made into flour, and is beaten into sun-dried pancakes.

Up to 100 ft. (30 m)

30–80 ft. (9–25 m)

The bark contains strong fibers, and these may be removed without causing the fatal damage that would happen to most trees. This tolerance permits the long-term exploitation of carefully managed trees. Mats, ropes, baskets, and even clothing items have been made with baobab fiber across a wide geographical range, making the tree important economically.

The leaves are widely eaten as a vegetable because they have a high protein content, which is remarkably balanced in amino acids. The leaves are also eaten for their mineral content.

Horse Chestnut

This deciduous tree is very familiar in northern temperate climates, especially in built-up areas or parks, but was originally a native of a very small area in the Balkans. Growing to 100 feet (30 m), with a considerable spread from branches that break out relatively low down on the trunk, it has long been popular as a shade and landscape tree, and is one of the few spectacularly flowering trees available for northerly latitudes. However, the palmate leaves are often disfigured by pests or diseases, with a leaf-miner moth being particularly problematic in recent years.

The white flowers are insect-pollinated, growing on an upright stem and giving the cluster a pyramidal appearance. They are followed by very distinctive, large, coarsely spiny fruit that break open to reveal nuts of up to 1¼ inches (3 cm) across.

- If sown soon after they ripen in fall and kept at ambient temperatures over winter, the nuts germinate readily the next spring. Growth is rapid.
- Young trees have some tolerance of moderate shade, but themselves cast increasingly dense shade as they grow, making it almost impossible to cultivate anything else below them.
- All soils, apart from very wet.

Up to 50 ft. (15 m)

Up to 100 ft. (30 m)

The nuts contain saponins, which have historically been used for washing fabrics and hair, although soft water is needed for a lather to form. They are still used in the manufacture of some shampoo brands.

The game of "conkers," whereby a nut on a string is flicked at an opponent's with the object of breaking it, originated in Britain but has spread worldwide—even adult competitions are now being held.

Research indicates that extract of horse chestnut, used as a dietary supplement, may be a safe treatment for circulatory problems such as chronic venous insufficiency (CVI) and varicose veins.

The forest here was almost composed of the kauri; and the largest trees from the parallelism of their sides, stood up like gigantic columns of wood.

CHARLES DARWIN, *THE VOYAGE OF THE BEAGLE* (1909)

Kauri

This evergreen conifer now grows only in the north of the North Island, New Zealand. Living for in excess of a thousand years, it can grow to become one of the world's largest trees. Having evolved in the Jurassic era, it, like other primitive conifers, still survives because of its ability to dominate its environment. Many adaptations have enabled its survival; for example, its constantly flaking bark prevents epiphytes from growing on its branches and so adding weight to them. Like that of another giant, the California redwood, its habitat has a long growing season and plentiful moisture, encouraging similar levels of growth. Biodiversity in both kinds of forest is very high.

One tree, known as Tane Mahuta, is currently the largest, at 167 feet (51 m) high and with a girth of nearly 59 feet (18 m). Other much larger trees existed in the past, reaching an incredible girth of 85 feet (26 m); fire has been a major cause of their destruction. In the past, occasional fires would have resulted in mass regeneration of the species, but the arrival of European settlers increased the frequency of these fires. Kauri forest is now protected from exploitation, and the tree enjoys a high cultural status as one of New Zealand's most distinctive species.

- Kauri seed should be sown immediately after harvesting because its viability declines rapidly after maturity.
- The tree prefers a well-drained but moist fertile soil, with a high humus content.
- Much of the relatively small rooting system is close to the ground surface, so generous mulching is worthwhile.
- Established trees survive several degrees of frost. They take approximately thirty years to grow to their full height.

Resin exuding from the tree lasts for centuries. During the latter half of the nineteenth century and the first of the twentieth, large quantities of semi-fossilized gum were dug up and used in varnish manufacture. That use has been superseded and today it serves only as an amberlike semiprecious stone in jewelry.

Kauri timber is of a very high quality, the best of today's limited supplies going into fine furniture and paneling. Historically the timber was favored for shipbuilding because of its resistance to decay, but rapid depletion of the forests led to a ban on logging in 1972 after a long campaign by conservationists.

The indigenous Maori people of New Zealand used dried Kauri sap as a form of chewing gum. The sap was also an ingredient of the dyes required for tattooing, which for centuries has been an important part of traditional Maori culture.

Kauris are a potential weapon in humankind's battle against global warming. It is one of the most effective absorbers of carbon dioxide, extracting far more than tropical rain forest. In this role it is equaled only by some species of eucalyptus, and the absorption is considered indefinite.

Up to 60 ft. (18 m)

Up to 165 ft. (50 m)

Tree of Heaven

This is a deciduous tree of great notoriety because it has proved to be an aggressively invasive species in many places where it has been introduced. A native of northeast and central China, it grows rapidly to 80 feet (25 m) and, if felled, it regenerates strongly. The leaves, up to 35 inches (90 cm) long, consist of up to forty-one leaflets.

Introduced to North America and Europe as an ornamental in the nineteenth century, the species soon began to spread by seed, aided by its production of allelopathic chemicals that suppress the growth of other plants. This enabled it to compete with even established native vegetation—although it is regarded as intolerant of shade. Even so, its ability to grow in depressed inner-city areas, where little else flourishes, has won it some friends and supporters.

- Being so vigorous and easily grown from seed, the species requires little guidance (or encouragement) for cultivation. Indeed it is more appropriate to discuss control. Felling must be followed by treating the stumps with herbicide immediately afterward to kill the stumps.
- The larger female trees, which produce high numbers of seeds, are the primary targets.
- For eradication, seedlings need to be hand-pulled before the taproot develops.

The timber is not of particular high-quality but has an attractive silky grain and is very flexible. This makes it the material of choice for making steamers—cooking implements of great importance in Chinese cuisine.

The species is the preferred food of the moth *Samia cynthia*, or Ailanthus silk moth, which has a long history in China as a source of strong, cheap silk. Being unable to absorb dyes, the material has limited uses, but it is still commercially important.

Up to 50 ft. (15 m)

The bark is the part normally used in traditional Eastern medical systems. Its astringent qualities assist in the treatment of dysentery.

Up to 80 ft. (25 m)

Albizia julibrissin ⚠

Persian Silk Tree

The small deciduous tree grows to 40 feet (12 m), with leaves divided twice into many very small leaflets. The flowers consist of white or pink clusters of stamens, and are very showy. The tree is commonly planted as an ornamental, both for the flowers and as a shade tree; a wide variety of insects are attracted to the flowers, as are hummingbirds. As with many pea-family species, the leaves close up at night. Noted as being allelopathic, the tree is potentially an invasive alien in some regions outside its natural range, which stretches from Iran to Japan.

Both the leaves and flowers are edible, although neither is especially tasty or nutritious. The foliage does make good animal forage, though, and the pealike fruits are attractive to livestock.

Several traditional Asian medical systems have used the bark or flowers as a sedative and an antidepressant; names such as "happiness herb" are used for it in several languages. The term "herbal Prozac" appears in English discussions about its properties.

The root nodules of Persian silk tree fix nitrogen in the soil. Considering that the tree is also useful for forage and shade, it has some potential in agroforestry and permaculture systems, particularly as a fast-growing pioneer tree in drier environments.

Up to 30 ft. (9 m)

Up to 40 ft. (12 m)

- The seeds should be soaked in hot water and left to stand for twenty-four hours. Even after this treatment it may be several months before they germinate.
- Young plants need protection from frost.
- Adult trees are hardy to -4°F (-20°C) but only in a continental climate; frost damage is more likely in a cool summer maritime one.
- The tree is notably tolerant of poor and saline soils, and of drought.

*Pumps, troughs, and boats were made
from alder, and bundles were commonly
used as a sort of retaining wall
alongside riverbanks.*

Alnus glutinosa ⚠

Common Alder.

Alder grows widely across Europe, and almost inevitably indicates wet soil or poor drainage. It grows to a maximum height of 100 feet (30 m), but the great majority are much shorter. Alders are often seen growing in colonies, which result from the ability of the species to throw up suckers, very often along the bank of a stream or lake. Their roots may grow into the water, where they form a sheltering habitat for young fish and other water life. The male flowers take the form of catkins, and are wind-pollinated. The tiny female flowers eventually form conelike structures, which then distribute large quantities of small seed to the winds. This level and type of seed production leads to it being a very effective colonizer of damp ground.

Alder has a symbiotic relationship with a bacterium on its roots, which fixes atmospheric nitrogen, enabling the tree to grow well in poor soils, adding to its ability to colonize. Growth is fast, more than 3 feet (1 m) a year for young trees. A wide variety of insects feed on the tree, making it important for biodiversity. A number of mushroom species have symbiotic relationships with the tree, in some cases growing in conjunction with it and no other species.

- Common alder seed needs either cold stratification or sowing in fall in order to germinate in the spring.
- Any moist, or even wet, soil is suitable, in full sunlight.
- The tree makes an attractive specimen, but it is more commonly planted as a quick windbreak.
- Alder's ability to colonize wet ground makes it very useful for restoration purposes, stabilizing banks and preventing erosion and flood damage.

A yellow dye may be extracted from the bark and young shoots, although it gives wool a reddish tone. With added copper a grayish-yellow dye is produced; this was used by tapestry weavers in Europe. Added iron compounds produce a black dye. A green dye can be extracted from the catkins.

Alder wood has long been popular for smoking fish, especially salmon, but also works well with beef and pork. It has a natural sweetness and mild flavor, which are readily imparted to the food and are not overpowering. As in all meat smoking, using damp wood will intensify the flavor.

Alder wood is soft and easily worked, which made it a popular material for making cheap items that required carving, such as bowls and clogs. Today, many such items are more easily made by molding plastic.

Up to 40 ft. (12 m)

The inner layer of the bark is the part of this tree that has been most used in traditional medicine; it is notably astringent, which has perhaps led to its popularity as an ingredient in mouthwash and toothpaste. It has also been used medicinally as a treatment for inflammation of the gums and the bleeding associated with it.

Up to 100 ft. (30 m)

Alnus rubra

Red Alder

This alder species has a relatively limited area of distribution: a narrow belt along the northwest coast of North America, but there it can be extremely abundant. Growing to a height of 50 feet (15 m), it is distinctive for the red-brown catkins in spring. Its natural habitat is along streams and in wetlands, often in association with willows and poplars. However, it is a pioneer species of open ground and so the clearing of woodland often results in high levels of seed germination away from wet areas. As a consequence, it is sometimes regarded as a weed tree by foresters seeking to replant the locally dominant conifer species. Because of its symbiotic relationship with nitrogen-fixing bacteria, however, it can help to increase nutrient levels in the soil for the benefit of other species.

- Red alder seed needs either cold stratification or sowing in fall in order to germinate in the spring.
- Any moist, or even wet, soil is suitable, in full sunlight or light shade.
- The tree is increasingly seen as a valuable species for rapidly growing biomass or timber, with trees being large enough to harvest in as little as twenty-five years.

Red alder was not highly rated traditionally, but it is now looked upon more favorably because it is easy to work and less expensive than most hardwoods. It was used by Native Americans for making masks and bowls. Today it features increasingly in construction and furniture making, for veneers, and for pulping. In furniture applications it can be necessary to redesign joints and the sizes of structural parts to compensate for the relatively low strength of red alder wood.

Up to 30 ft. (9 m)

Up to 50 ft. (15 m)

Because it has a close grain and readily accepts staining, red alder wood may be used to imitate cherry, mahogany, and even walnut. The roots have long been used for basket making by Native American tribes, who also derived a red dye from the bark.

The bark was used by Native Americans to treat skin irritation, including that caused by poison ivy and poison oak, and digestive complaints. The bark contains salicin, the precursor to aspirin.

Alstonia scholaris

Indian Devil Tree

Also called the blackboard tree and the milkwood pine, the Indian devil tree is an evergreen species of humid or seasonally dry environments in the Indian subcontinent, Southeast Asia, and parts of northern Australia. It can be a major element in lowland forest, particularly in storm-prone areas. The glossy oval leaves are arranged in whorls around the twigs. The tree may grow to 130 feet (40m), and develops a very even shape with branches also arranged in whorls, making it an attractive species for urban and park planting. Branches may be shed in severe winds, but they then regenerate from the trunk. A milky-white latexlike sap exudes from cut surfaces—a common feature of the Apocynaceae family. The name Indian devil tree reflects popular prejudices that it contains evil spirits, possibly a consequence of it being poisonous. In parts of India, people are reluctant to sit or pass under the tree for fear of the devil.

The bark is used in traditional medical systems against malaria, skin conditions, and digestive disorders such as dysentery. It has been researched as an anti-cancer drug, but it has no current use in evidence-based medicine.

There is some, albeit slim, evidence that the seeds have been used as a psychoactive drug for Aboriginal ceremonies in Queensland, Australia, and in Tantric rituals.

Indian devil tree wood is notably soft and in many traditional Asian cultures has been used for making pencils or writing tablets—hence the common name of "blackboard tree"—as well as for paper making. In Indonesia it is also considered suitable for making coffins.

- Seeds germinate best if soaked in water preheated to 122°F (50°C) for thirty minutes before sowing.
- Young plants grow rapidly in full sunlight and are adaptable to most types of soil.
- The Indian devil tree does not tolerate frost.
- Planting is especially recommended in locations where storms and cyclones are frequent for its regenerative capability.

Up to 65 ft. (20 m)

Up to 130 ft. (40 m)

India earns millions of dollars a year by exporting cashew kernels.

"CULTIVATING CASHEW NUTS" ON DEPARTMENT OF AGRICULTURE (SOUTH AFRICA) WEBSITE

Cashew

The cashew tree is an evergreen species from northeastern Brazil. It can grow to 46 feet (14 m), and its seeds are of major culinary and economic importance. The small reddish flowers, which grow in clusters, are each succeeded by a pear-shaped "apple" (technically it is not a true fruit as its anatomical origin is different). Hanging below the fruit, in an almost comical manner, is a large, curved, single seed—the cashew "nut"; the edible kernel is contained inside.

Cashew cultivation, once only of marginal value, is becoming more important because the nuts are an ingredient in a number of cuisines of increasing popularity. There are also many industrial uses for by-products. Nigeria and India are the largest producers. Trees usually start to produce economically significant quantities of nuts at eight years of age; recent breeding has produced smaller-growing and more efficient trees that can produce worthwhile nut yields in three years.

One particularly famous cashew tree in Brazil has branches that take root wherever they touch the ground. This habit has enabled the tree to occupy around 1¼ acres (0.5 ha) of land.

- Cashew nuts germinate rapidly in warm, moist conditions, and trees establish easily and rapidly. Even so, the more commercial, better-quality varieties are typically propagated by grafting.
- A tropical climate with a minimum rainfall of 40 inches (100 cm) is required (even more rainfall is ideal).
- A monthly mean temperature of 77°F (25°C) is also essential.

The cashew "apple" is easily bruised and so rarely sees distribution beyond local markets, where it is appreciated as a very juicy and sweet fruit with a sour edge. It is commonly processed into a refreshing, astringent drink, and is also distilled as a liqueur.

Cardanol, a chemical derived from the anacardic acid in the shell, is used in the manufacture of resins, coatings, stabilizers, paints, and plastics. Its chemistry gives it great potential for being the start of a number of pathways for synthesizing other compounds.

A well-known snack, cashew nuts contain a higher proportion of starch than most nuts. This makes them useful as a thickener for sauces or stews in a number of warm-climate cuisines. They are often used as a basis for desserts, chiefly in India and southern Africa.

The oils contained in the cashew shell are highly irritant because they contain a compound very similar to that found in poison ivy. This characteristic makes it necessary for those handling the unprocessed nuts and fruit to wear collective clothing. The nuts have to be roasted to remove the skin, a dangerous procedure to operatives because the smoke is highly toxic—consequently, this stage of cashew processing always takes place in the open air.

Up to 50 ft. (15 m)

Up to 46 ft. (14 m)

Agarwood

In the world of commerce, agarwood is a rather strange commodity because the tree only has economic value if it is infected by a fungus— when it then becomes very valuable. A large evergreen tree native to India and Southeast Asia, it normally has pale-colored wood. But if it becomes infected with a particular fungus, *Phaeoacremonium parasitica*, it produces large quantities of resin in an attempt to suppress the infection. The result is a dark-colored and heavily aromatic wood that has been sought after for centuries; its history of over-exploitation has resulted in a "vulnerable" rating by the CITES conservation authority.

Several different species in the *Aquilaria* genus produce agarwood, but *A. malaccensis* has been the one traditionally preferred by traders. Research currently being carried out is aimed at identifying alternative sources of agarwood, discovering sustainable harvesting methods, and finding ways of including the tree in agroforestry.

- Seed loses viability within weeks of maturity.
- Agarwood trees are easy to grow, however, and are commonly grown in parts of northeast India.
- The trees need little care and are tolerant of a wide range of conditions.
- Traditionally, bark was peeled off trees to facilitate the desired fungal infection.

Agarwood oil is the most expensive essential oil in existence, and is in particularly high demand in the Arab world, where it has long been used for perfume, and in Japan, where it is chiefly used as an ingredient of incense.

Marked with dark streaks, agarwood has always been popular for carving small and valuable items, such as those commonly worn as jewelry. It has also been a favored material in Asia for making carvings of the Buddha.

Up to 10 ft. (3 m)

Renowned for its deep and complex odor, rather than any therapeutic benefits it might have (although it is claimed as an aphrodisiac), the essential oil derived from agarwood is used in various high-value cosmetics.

Up to 130 ft. (40 m)

Araucaria bidwillii

Bunya Pine

A large, imposing conifer, the bunya pine is truly a living fossil, the last remnant of its section of the ancient *Araucaria* genus. During the Mesozoic era (252–66 million years ago), its relatives covered vast areas of the globe. Growing to 150 feet (45 m), with very tough dark green leaves, it has a distinctly prehistoric appearance. Native only to a few disjunct areas in Queensland, Australia, it was distributed early on by the continent's European colonists; consequently, fine specimens may now be seen in New South Wales and Western Australia, too.

The football-sized cones fall from the trees without splitting open to release their seeds, causing speculation that they evolved to be eaten (and thus dispersed) by dinosaurs or large, now extinct, mammals. The size of the crop varies from year to year; in the past, Aborigines held feasts when harvests were good, with different tribal groups traveling great distances to come together and feast on the large nuts held within the cones.

- Fresh seed germinates within six months at a temperature of around 59°F (15°C).
- Seedlings should be planted out into their intended growing positions as soon as possible.
- Established trees are hardy to around 23°F (-5°C), or possibly as much as 18°F (-8°C).
- The trees are tolerant of most well-drained soils and show some tolerance of salt spray.
- A good bird habitat for large gardens.

Bunya pines were held in high regard by the Aborigines, and family links to particular trees were among the few examples of personal ownership by presettlement peoples. The Aborigines ground the nuts into flour and used it to make a kind of bread, and also fermented it. Today the nuts are recognized as nutritious and delicious, and are increasingly being grown and promoted as a food crop.

Up to 80 ft. (25 m)

Up to 150 ft. (45 m)

Earlier generations of Australian settlers used bunya wood for quality cabinet work, and paneling. Today the wood is in short supply, trees being authorized for removal only when they are in poor health or for safety reasons. Small plantations are being established.

The timber's main modern use is as a tonewood for acoustic guitar soundboards—it is probably Australia's best wood for the purpose. Bunya is a sustainable wood because it attains maturity in eighty years, rather than the 300-plus years required by spruce.

Areca catechu

Betel Palm

The betel palm is found wild in Southeast Asia, southern India, and parts of East Africa. It grows to a maximum height of 65 feet (20 m), with pinnate leaves up to 6½ feet (2 m) long. Primarily known as a source of betel nut, which is widely chewed as a drug, the tree has a neat appearance and adaptability that have led to its use as an atrium plant in buildings in the temperate zone.

The palm is extensively grown in some Southeast Asian countries and India for nut production. The trees have a long commercial life, up to a hundred years. Harvesting is carried out annually, with bunches of nuts being cut when ripe, and then sun-dried. Many nuts are exported because betel chewing is popular in places outside the area where the palm can be grown, such as parts of China.

- Betel palms have occasionally been grown by palm enthusiasts, although obtaining nuts outside the tropics is difficult.
- Fresh nuts germinate rapidly at 77°F (25°C).
- Warm temperatures and full sunlight are needed.
- Palms may be grown under cover, but these have increased vulnerability to various insect pests.
- Palms usually begin to fruit after seven years.

Betel nuts contain a number of psychoactive substances, act as a stimulant, and are possibly slightly addictive. It is suggested that around 10 percent of the world's population regularly chews betel, usually in the form of a "quid" containing shavings of the nut, some lime, and a wrapping made from a leaf sourced from a climbing plant related to pepper.

Up to 46 ft. (14 m)

Up to 65 ft. (20 m)

! Betel-nut chewing is associated with serious health risks, including mouth ulcers, gum disease, oral cancers, stomach ulcers, heart disease, and addiction.

Persian descriptions of betel chewing appeared in Indian literature of the eighth and ninth centuries.

DAWN F. ROONEY, "BETEL IN SOUTHEAST ASIA" (1995)

Argan

This spiny, compact tree, which grows to around 25 feet (8 m) and looks like any other tough dry-country tree, is now renowned as the source of a highly valued oil. Native to a limited semi-desert area of Morocco and Algeria, culturally largely Berber, the tree used to be traditionally managed for a variety of products, but cutting for firewood and agricultural changes have resulted in large losses. In 1998 the surviving area of the so-called arganeraie was designated a UNESCO Biosphere Reserve.

The fruit, 1.5 inches (4 cm) long, contains pulp, and within that a hard nut that contains one or more seeds—these are the source of the oil. The value placed on the oil is now encouraging conservation, as well as new plantings of the tree in the Middle East and the United States.

- Fresh seed germinates relatively well after soaking in water, but relatively few seedlings may survive.
- Propagation from cuttings taken from young growth is also possible.
- Irrigation is needed to help trees establish, as well as protection from animals.
- Once trees are well established, little needs to be done; however, as with any desert plant, good growing conditions in the form of water and feeding will promote a higher rate of growth.

Traditionally, argan oil was used in a similar way to olive oil in neighboring regions around the Mediterranean: for dipping bread, cooking, and preparing savory spreads.

High-density argan wood was formerly used for house construction. Now the wood's main use is in quality jewelry and souvenirs. Being dense, it also makes very good charcoal.

Argan oil is full of minerals and antioxidants, leading to claims that it is exceptionally useful for nourishing and reviving skin and hair. While Berber women use the oil neat, in the industrialized world it is generally incorporated into a range of cosmetic products.

Argan oil has been a component of Berber folk medicine for centuries.

MATTHEW WILSON,
"MOROCCAN ARGAN OIL,"
FINANCIAL TIMES WEBSITE

Up to 12 ft. (3.5 m)

Up to 25 ft. (8 m)

Jackfruit

Growing to 80 feet (25 m), this evergreen tree from the southwest of India bears the largest edible fruit of any tree, 4 feet (1.2 m) long and weighing up to 80 pounds (35 kg). The tree plays an important part in small-scale agriculture and in agroforestry. The leathery-textured leaves are distinctively lobed when young but become entire later.

As a tree of many uses, it has been cultivated for millennia in India and was distributed throughout Southeast Asia at an early date; further distribution occurred during the nineteenth century. Occasionally the species has become an invasive alien, as in, for example, around Rio de Janeiro, Brazil. A number of cultivars exist, having been selected for their superior productivity or particular characteristics.

Until now, jackfruit has had relatively low status and has not been extensively exploited commercially; however, it is now being suggested by researchers that the tree could be a good alternative to grain crops in areas affected by climate change. Unlike wheat and corn, the tree can cope with high temperatures with no loss of yield. The Indian central government and some state governments have started to promote the tree, as well as the processing industries that can utilize it.

- Fresh seed germinates over several weeks at temperatures of around 77°F (25°C).
- Anyone in the tropics seeking to raise jackfruit is best advised to use named varieties grown from cuttings or grafts.
- Consistently high temperatures with plentiful moisture are needed for growth, although the trees can survive short periods of near-freezing temperatures and will tolerate droughts of up to four months long.

Popular as a fruit, jackfruit is eaten fresh, processed into juice, or sometimes sold chipped and dried, to be cooked and eaten like potato fries. The flavor is sweet but relatively subtle. Jackfruit is a high-calorie foodstuff, so has the potential to play an increasingly important role as a staple food, particularly in the context of smallholder agriculture.

Jackfruit trees are traditionally grown for timber in Sri Lanka and parts of Indonesia because the wood is of high quality with good resistance to termites and fungal decay. Accordingly, it has been a favored wood for building royal palaces. Somewhat resembling mahogany in appearance, it is more comparable to teak in its physical characteristics.

Up to 50 ft. (15 m)

Up to 80 ft. (25 m)

The seeds, each ¾ to 1½ inch (2–4 cm) long, are prepared in the same way as nuts, and when roasted their flavor resembles that of chestnuts. They can be also be boiled and then peeled. The seeds are popular for snacking and as an accompaniment to a meal.

Unripe jackfruit has a flavor oddly reminiscent of meat, which has led to it being used as a meat substitute in a range of spicy dishes, particularly in India, where a large proportion of the population is vegetarian.

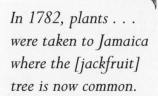

In 1782, plants . . . were taken to Jamaica where the [jackfruit] tree is now common.

JULIA MORTON, *FRUITS OF WARM CLIMATES* (1989)

Azadirachta indica

Neem

For a tree that survives desert conditions, the neem looks surprisingly lush, thanks to its ability to tap deep reserves of groundwater. Growing to 100 feet (30 m), the deciduous tree has been spread, largely by Indian diaspora communities, far beyond its original homeland of the northern Indian subcontinent and Southeast Asia. In hot, dry conditions the shade cast by its mass of feathery, pinnate leaves is greatly welcomed by both people and livestock.

Neem has long been famed for its wide range of uses, which are based on its chemistry. Oil from the crushed seed is widely used in the manufacture of cosmetics and a range of medical products, chiefly in India. There are also some modern industrial uses for it; for example, it has a role in making plastics. There has been much interest from pharmaceutical companies, too; a long-standing battle between one company and the government of India over the right to patent certain uses of the tree concluded with the case being won by India. Some campaigning groups were especially critical of attempts by corporations to profit from what country people in India have known about the tree for generations, describing their actions as "biopiracy."

- Neem seed seems to lose viability very quickly, so collecting fresh seed is vital for germination.
- Growing seedlings presents few problems, although the plants cannot tolerate freezing temperatures or prolonged cold, and are sensitive to waterlogging.
- Trees grow best where rainfall is 16 inches to 4 feet (40–120 cm) per year, and tolerate high temperatures when established.

Neem is one of the most widely used, easily available, and easily prepared natural insecticides in India. Rather than being directly toxic, it works by inhibiting or suppressing reproduction. Neem products can be used on agricultural pests, animals, and on human head lice without risk to humans.

Certain constituents of neem have been shown to function as contraceptives (for women) and have been marketed in India. The bactericidal and antiseptic properties of neem are also well researched, and products such as neem "toothbrushes" or chewing sticks are widely used in the Indian subcontinent.

Young neem leaves and shoots are used in a number of Indian and Southeast Asian cuisines, not so much for their flavor (which is bitter) but because it is believed that they are healthy. Other ingredients counteract the bitterness.

People in India scatter neem leaves among their stored clothes to repel insects such as moths and beetles. The leaves are also wrapped in cloth and immersed in sacks of rice and other stored food products to help prevent insect infestations.

Up to 100 ft. (30 m)

Up to 100 ft. (30 m)

Betula spp.

Birches

Very familiar to people living at the northern latitudes of North America or Eurasia, birches also grow in the foothills of the Himalayas and associated mountain ranges. Birch species tend to be very similar and botanists have varied opinions on the division of the *Betula* genus—as many as sixty or as few as thirty species have been suggested. Birches are short-lived pioneer species, found in challenging environments, and rarely grow to more than 65 feet (20 m), usually very much less. The European species are a particular feature of abandoned industrial sites. All tolerate cold and grow well in low-fertility soils and in short growing seasons. Except in very cold climates, they tend to be displaced by longer-lived tree species.

Birches stand out particularly well because of the distinctive bark—that of most of the northern species is pure white, popularly dubbed "silver." Others, particularly Asian species, may have gray or pink-toned bark. The bark is also liable to peel and with age becomes increasingly rugged. The look of birch bark and the tree's adaptability have ensured widespread use by the landscaping industry.

- Birch seed is best sown fresh, in fall, so that it chills over the winter.
- Germination tends to be rapid, as is subsequent growth. Young trees are best planted in an open situation where the soil is likely to stay moist and cool, but also to be well-drained.
- Birches react badly to attempts to prune or shape them, and realistically can only be grown to follow their own natural shape.

Birch is well-thought of as a wood, being both attractive in appearance, lightweight, and durable. It makes particularly high-quality plywood, which is favored by furniture and toy makers because it does not crack or splinter easily. Its rigidity and the way it responds well to sound makes it popular for making drums and speaker cabinets. Birch bark has been used in the past for making portable containers in which to collect food or water by both Native Americans and Eurasian country people.

Birch burns well, even when "green" or freshly cut, and so does not require drying before use; finely split, it makes good kindling. Of all woods for burning, black birch (*B. lenta*) probably burns hottest and longest.

Up to 50 ft. (15 m)

Up to 65 ft. (20 m)

The Sanskrit name for birch is *bhurj*, which comes from the same Indo-European root word as the English name. The sacred Indian language of Sanskrit has a close association with the tree because the bark was used for the recording of sacred texts in Vedic times (c.1750–500 BCE).

Birch trees may be tapped for their sap in the spring, a traditional Native American practice. The sap is slightly sweet and may be fermented to make wine, or concentrated by boiling, but its flavor does not compare to that of maple syrup.

For the celebration of their victories in combats it was customary for the Nayars to consume intoxicating liquor. They therefore coerced the Ezhavas to supply them with palm wine for they were too indolent [for] manual labor.

KANIPPAYYUR SANKARAN NAMBOODIRIPAD, HISTORIAN

Borassus flabellifer

Asian Palmyra Palm

With an almost spherical head of massive fan-shaped leaves, and growing to 100 feet (30 m), the palmyra or toddy palm is a distinctive sight on the skylines of India and Southeast Asia. It has five other relatives in the genus that are very similar, which are found in New Guinea, Madagascar, and humid tropical regions of Africa. At any time, a mature tree may have between twenty-five and forty living leaves, each one of them up to 10 feet (3 m) across and edged with sharp spines.

Among the best known specimens are the ones growing around the great temple complex of Angkor Wat in Cambodia, where the palm is almost a national symbol. That is true also of the Indian state of Tamil Nadu, where the tree is known as the "celestial palm" because every single part of it can be used for something. This usefulness has resulted in its very extensive cultivation for millennia.

The fruits are borne in bunches and resemble coconuts, but are smaller with harder contents. Before they have fully ripened, the fruits contain a translucent, jellylike liquid containing the seeds, as well as a clear liquid that is similar to coconut milk.

- The nuts germinate in about a month at 86°F (30°C) and send out a long taproot, so deep pots are needed.
- They should be planted out reasonably quickly, but will only survive in continually warm conditions.
- When established they will tolerate dry seasons, but need plentiful moisture in between.
- Growth is not fast and they tend not to fruit until they are at least twenty years old.

The fruit can be eaten, both the ripe kernel and the soft contents of the unripe one. The milk is used as an ingredient in making treats in southern India. The seeds are also eaten when they are germinating and producing fleshy stems. The seeds are starchy and very nutritious, and can be prepared in a number of different ways as a savory dish.

The leaves, suitably prepared, by boiling and then polishing the surface, have long been used to write on—historically, the texts were mostly religious. The writing surface splits easily, so southern Indian writing styles, like Malayalam and Tamil, were developed to be cursive, with a strongly horizontal character, so that the writing implement would not pierce the leaf.

Borassus flabellifer is just one of many palms producing sap that can be fermented over the course of a single day to make an alcoholic drink, known as toddy. The "wine" is extremely cheap to make and is widely drunk in both urban and rural areas.

The long leaves of the toddy palm are primarily used in traditional house construction wherever the palm grows. Leaf segments, if attached to a suitable support matrix, function very effectively as waterproof and very durable roofing tiles. The fibrous leaves are also used to weave baskets, hats, mats, and umbrellas against the rain and sun.

Up to 50 ft. (15 m)

Up to 100 ft. (30 m)

Broussonetia papyrifera syn. *Morus papyrifera*

Paper Mulberry

Reaching 65 feet (20 m), and sometimes having a shrubby habit, this deciduous species originated in the warmer parts of East Asia and Southeast Asia, but was long ago distributed more widely over Asia and the Pacific region, and then farther afield still during the colonial era. The tree's rough, hairy leaves with their varied shapes make a connection to true mulberries, and in fact in the past it was classified as a *Morus* species. Male and female flowers grow on separate plants. The females produce spherical orange edible fruit, which is eaten by a wide range of wildlife, thus helping to spread the plant. It is wind-pollinated. The tree also suckers, particularly if felled, so thickets can build up.

Paper mulberry can grow quickly and in the past has been planted as a shade tree in cities in warm-climate zones, where the tree's ability to withstand air pollution, as well as recover quickly from damage, stands in its favor. When the tree is grown for paper production, it is usually coppiced to maximize its production of new bark.

- The seed, once it has been washed free of the fruit, germinates easily, and the seedlings grow rapidly.
- Trees need full sunlight but seem able to accommodate most types of soil.
- The paper mulberry is hardy to around 14°F (-10°C) but does not seed in cooler-climate zones, a factor that naturally limits its potential for invasiveness there.

The leaves can be eaten, or used to wrap food for steaming. They can also be used to feed silkworms, but other mulberry species are preferred and more successful for that purpose.

The leaf juice is used in preparations to treat skin conditions and insect bites. The juice is also used internally as a diuretic, laxative, and for promoting perspiration.

The paper mulberry's name states why the tree has been so widely cultivated. The inner bark contains fibers that, once removed by soaking and steaming, can be made into a paperlike material by a process of pounding with wooden mallets. The more it is pounded, the softer and more flexible the material becomes, even to the point of resembling a cloth. This fabric was used to make clothing until well into the twentieth century on some Pacific islands, and is still used for ceremonial occasions. As a paper it has had a huge range of applications in East Asia.

There is little doubt that this tree's remarkable capacity to spread has led to it becoming a major "invasive alien" in warm-climate areas. Its seed is spread by birds into wasteland, which it colonizes very successfully. The tree is a nuisance in another way, too—it sheds large quantities of pollen, and the pollen is notably allergenic, contributing to the misery of hay fever sufferers wherever it grows.

Up to 80 ft. (25 m)

Up to 65 ft. (20 m)

Mulberry paper was invented and first used around 105 CE in China by Ts'ai Lun, as documented in his announcement to Emperor Hi-Ti.

"A SHORT HISTORY OF MULBERRY PAPER," ASIA CREATIONS WEBSITE

Bursera simaruba

Gumbo-limbo

This large tree of the Caribbean, northern South America, and the very southern tip of Florida grows to 40 feet (12 m) and is commonly found on drier soils, including sandy and shallow, calcareous ones (containing lime or chalk). The tree is notably tolerant of salt spray, and so can be an important presence in coastal forest communities. It is deciduous and may drop leaves in the dry season.

The leaves are pinnate and the bark is red, resinous, and rather thin, with the look of having been recently varnished; the bark may also peel. Branches tend to droop down as the tree ages, and there is also a tendency for multiple trunks to develop.

Although gumbo-limbo has not been used much historically as a decorative species, its attractive appearance and tolerance of difficult conditions have led to an increase in its use in landscaping. The seeds are eaten avidly by a variety of bird species and the tree has an important role in fostering biodiversity in urban environments.

- Freshly sown seed germinates quickly at 68–77°F (20–25°C), as do cuttings.
- Larger branches thrust into the ground also root very easily, although they tend not to develop into well-shaped trees.
- Does not tolerate frost or prolonged cold, but can withstand drought, making irrigation unnecessary.

A resin smelling of turpentine exudes from cuts in the surface of the trunk. It has anti-inflammatory qualities and has been used for treating gout; usefully, it helps to counteract irritation caused by *Metopium toxiferum* (poisonwood), which often grows alongside.

Because the branches root easily and the tree is fast-growing, gumbo-limbo is widely planted by rural people across its range as a cheaply produced yet sturdy form of "living fence."

Up to 40 ft. (12 m)

Up to 40 ft. (12 m)

The wood is soft, light, and easily worked. Traditionally, it was popular for carving fairground carousel animals, before the arrival of molded plastic. It is also used for matchsticks, packing material, and furniture.

Buxus sempervirens

Box

Box is an evergreen tree or large shrub known to most people as a closely clipped feature in European or North American gardens, where it appears as either low, ornamental hedging or sculpted into geometrical or figurative shapes; increasingly, clipped single box trees are grown in freestanding planters as well. Box topiary has been popular since Roman times, and, like yew topiary, it is an important feature in many historic gardens. Left to develop naturally, however, the tree can grow to 30 feet (9 m) high, but trees of this size are rarely seen. A native of dry, alkaline soils across Europe, box shows considerable drought tolerance, which is an additional reason for its popularity in cultivation, especially in the Mediterranean region. A great many forms have been selected for cultivation, particularly in the American South, on the basis of their leaf shape, overall plant shape, leaf coloration and texture, growth rate, and cold hardiness.

Boxwood has the distinction of being Europe's densest wood. Fine-grained and lacking growth rings, it has been used for small, precious objects, including musical instruments and boxes, and handles and heads for specialist tools.

Small-leaved box responds well to clipping, hence its popularity for topiary. Since the late medieval period in Europe, growing low box hedging in decorative patterns as parterres has been revived repeatedly in fashionable gardening.

• Box is nearly always grown from cuttings, which keep the distinct character of the parent plant.
• Box is easy to grow in any well-drained soil. Although it tends to be grown in soils with high fertility, it arguably stays healthier (although grows more slowly) on the calcareous soils of lower fertility on which it is found in nature.
• Clipping into neat hedges or topiaried shapes is best done during the growing season.
• Shaded areas of box hedges may not thrive as well as areas growing in full sun.

Up to 25 ft. (8 m)

Up to 30 ft. (9 m)

The wood makes some of the best engraving blocks, a use made famous by the English artist and natural history author Thomas Bewick (1753–1828), who employed the tools of metal engravers to carve against the grain and secure very fine detail.

BOX | 047

Pau Brasil, Brazilwood

Of all the trees that have been ruthlessly exploited by humanity, this is perhaps one that has suffered the greatest depredation. An evergreen, growing to no more than 50 feet (15 m), it is not large, but being a climax species it does not establish easily in open conditions. This has hindered attempts to grow it commercially or to re-establish it in areas where it was once common.

Commercial exploitation of the tree began as soon as it was discovered, and its timber was regarded as so valuable that ships carrying it were liable to be attacked by pirates. Given its distinctiveness and importance, the region where it grew—later to become the country of Brazil—was named after it. "Pau brasil" derives from a name which means "emberlike stick" in Portuguese, a reference to the deep red of the wood.

- Seed germinates readily at warm temperatures, and the trees make reasonably good growth in fertile, alkaline soils in full sunlight.
- Pau brasil tends not to perform well unless it is an integrated part of a forest community.
- Unfortunately, the all-important heartwood from plantation trees is not of the same quality as is found in the wild-harvested ones.

The reddish-toned wood is very dense, and can be polished to produce an exceptionally high luster. It became sought after for use in marquetry, veneers, and the production of high-value items, in particular for the making of bows for violins and cellos, for which no other wood is as good, both acoustically and in terms of tensile strength.

Reduced to sawdust, Pau brasil can be made into a powerful red dye. When discovered, this was one of the strongest red dyes known, until it was replaced by synthetic dyes in the nineteenth century.

Up to 40 ft. (12 m)

Up to 50 ft. (15 m)

Guayacaú Negro

The guayacaú negro, or black guayacau, is an example of a relatively obscure species that has been under exploited but could have great potential. Growing to around 60 feet (18 m) tall, it has finely divided evergreen pinnate leaves and yellow flowers, which are a good nectar source and very attractive to honey bees. The tree is native to the Chaco region of the Paraguay, Bolivia, and Argentina borders, a vast almost featureless low forest that has a long and very hot dry season. This area is undergoing considerable deforestation due to the expansion of cattle ranching, and consequently the species is listed as "vulnerable." Like all the species of the *Caesalpinia* genus, it has attractive flowers and foliage and has been promoted as a landscape tree, for which it certainly has potential.

The wood is an attractive dark brown color and has a good reputation for hardness. As a timber tree it is regarded as being largely under-exploited, and so it faces an uncertain future.

The tree's potential in agroforestry is largely as a forage plant, because the pod pulp is palatable and nutritious for livestock. Unusually, pods are produced over a long period and dropped over a longer one, in effect for most of the year.

Research indicates that the tree has anti-inflammatory and anti-microbial properties; it may have potential for use against bacteria that have become resistant to antibiotic drugs.

- Seeds need to be soaked for twenty-four hours; any that fail to swell up should be filed before re-soaking.
- Germination is likely to be rapid.
- Little is known about the onward cultivation of this tree, but it is likely to thrive in any frost-free, seasonally dry climate.
- There is likely to be a tolerance of the drought and poor soils that characterize its native habitat.

Up to 60 ft. (18 m)

Up to 60 ft. (18 m)

Papaya

The papaya may only doubtfully be defined as a true tree. Growing to around 35 feet (10 m) at the most, and rarely living for more than twenty years, papayas tend to form a single, non-woody trunk, with a mass of leaves emerging from the top. Female flowers give way to distinctive elongated fruit, which may grow as long as 18 inches (45 cm) and become dull orange in color when ripe.

Native to Central America, the papaya plant is thought to have been taken into cultivation several millennia ago. The Spanish colonists carried it to many corners of their American empire, and from there it has been spread to the rest of the world. As well as being a major commercial crop with uses other than as a food, it is also an important part of many smallholder farming systems.

- Papayas do not transplant well, so seed needs to be sown where the plants are to be cultivated, or started off in pots.
- Germination and consequent growth is rapid, and plants can produce fruit in as little as a year.
- To thrive, papaya needs plentiful moisture and very fertile soil in full sun in tropical or sub-tropical conditions.
- Plants will not survive prolonged cold or hard frost.

Papaya leaf has traditionally been used to treat malaria and dengue fever, and as a contraceptive. It also has potential use in treating wounds and intestinal parasites. The fruit is currently being promoted as having a wide range of health benefits.

Papain is a compound found in green fruit and in the latexlike juice of the plant. It has long been used as a tenderizer for meat and other tough, fibrous proteins, and is often used commercially.

Up to 20 ft. (6 m)

Up to 35 ft. (10 m)

Saguaro

Immortalized as an image of the American West, this giant cactus can grow to 65 feet (20 m). It is native to southwestern North America, and other very similar species are found elsewhere, such as the even larger *Pachycereus pringlei* of northern Mexico and *Echinopsis atacamensis* of northeastern Argentina (both known as cardons). Assisted by a symbiotic relationship with nitrogen-fixing bacteria, they are able to grow in extremely infertile and dry environments.

The plants grow slowly, and after many years usually develop side-branches. The white flowers open at night and are surprisingly lush-looking for a desert plant. Inside the plant is supportive tissue that is surprisingly strong despite being full of holes.

Saguaro fruit, like that of all cactus, can be eaten. It was certainly important for some Native American tribes, but now its consumption is largely historical.

Saguaro and cardon timber has been used to make furniture, gates, small structures, and even church roofs. It is highly decorative and is still used for making high-value items. Much of it has been collected from dead plants in the wild, which is possible because it decays very slowly in the desert. However, living plants are often protected.

- Seed germinates easily in favorable conditions, taking only a number of weeks.
- Young plants need to be grown on in full sunlight and require careful watering.
- The growing conditions of saguaro are much like that of any other cactus—soil drainage needs to be perfect, and if the plants are likely to be subjected to frost they must be kept as dry as possible.
- Growth is slow, but the actual rate is dependent on what level of soil moisture is maintained.
- The first flowers appear when the plant is about thirty-five years old.

Up to 20 ft. (6 m)

Up to 65 ft. (20 m)

> *[It] sends out a massive mazelike array of roots very close to the surface.*
>
> U.S. NATIONAL PARK SERVICE, ARIZONA WEBSITE

Hornbeam

A major component of deciduous woodland across Europe, as well as parts of Turkey and Iran, the hornbeam grows to 80 feet (25m) and is familiar as a tree in planted landscapes, especially as a hedge or street tree. The leaves are superficially similar to those of beech—an unrelated genus—although more deeply pleated with distinct teeth; they turn a rich gold in the fall. The trunk has a very distinctive fluted profile and becomes ridged with age. The fruit has wings, which is perhaps the most immediately recognizable feature for the tree's identification. Hornbeam's use in gardens and landscapes has led to the selection of a number of cultivars, often for their distinctive upwardly swept branching habit. Hornbeam, like many forest trees, has roots that form linkages with a range of mycorrhizal fungi.

- Seed needs four months of stratification, and even then it may germinate only in the second spring after sowing.
- Seed collected before it is fully ripe has a better chance of germinating in the first spring after sowing.
- The tree is tolerant of a wide range of soils, and often does better than beech on heavy clay. It is cold tolerant but does best in warm-summer climates.

One of Europe's hardest woods, hornbeam can be too hard for use as normal timber. It has been used for making tool handles, gearwheels in windmills and other machinery, wooden screws, and shock-resistant components.

The inner bark and leaves were traditionally used to make preparations for treating rheumatism and toothache and soothing sore muscles. A compress of leaves can help to stop bleeding.

Being very dense, hornbeam makes very good firewood, burning at the highest temperature reached by a native European wood. Consequently, it was used extensively for charcoal.

Up to 50 ft. (15 m)

Up to 80 ft. (25 m)

Carya ovata

Shagbark Hickory

One of several species of hickory that make up an important part of the mature forest communities of eastern North America, shagbark hickory grows to 80 feet (25 m). Its large pinnate leaves—up to 2 feet (60 cm) long—are typical of the entire walnut family. But the tree's distinguishing characteristic is its shaggy bark, as indicated by the common name.

The nuts are edible with a sweet flavor and are an important food source for a wide range of wildlife, just as they once were for Native American tribes. They are similar to the closely related pecan. The trees are unreliable nut bearers, so the species has never been commercially developed. The hickory is also rather slow growing, which has limited planting, even within its natural habitat.

- The large seeds need to be sown in fall or else stratified before sowing.
- Seedling growth appears to be slow for several years, but actually the plant is building up an extensive root system— which accounts for hickories' notorious dislike of being transplanted.
- Deep, fertile, and moist soils are preferred.
- Trees recover well from coppicing.

All hickories burn well, and they are favored for smoking meat, fish, and other foods because of the rich flavor that they impart. Chipped hickory wood is available for finessing the smoke produced by home barbecues.

Hickory wood is hard, dense, and heavy, making it suitable for the handles of tools that receive shock and stress. Native Americans traditionally favored it for making powerful bows.

Enjoyment of shagbark hickory nuts is, for people who do not live in forested areas, limited to a chance discovery of a tree during a hike in the wild. The nuts' rarity is due to their not being worth growing commercially. The trees do not start bearing until they have grown for forty years, and then they only bear fruit every three to five years.

Up to 50 ft. (15 m)

Up to 80 ft. (25 m)

Castanea dentata

American Chestnut

Few trees have had a history as tragic as that of the American chestnut. Once upon a time, it dominated huge areas of the eastern regions of North America and comprised up to 25 percent of the forest biomass of the Appalachians. A deciduous tree, capable of growing to 100 feet (30 m), it played a major ecological role in the region, producing nuts that fed wildlife and nutrient-rich foliage that recycled the nutrients of the forest floor. It was also economically very important.

In 1904, however, imported chestnut trees from Asia introduced a fungal disease, known as chestnut blight, into the United States, and within a few decades the tree was practically wiped out. Although the stumps survive, newly sprouted trees become reinfected when they reach a particular size. Efforts are therefore being made to develop disease-resistant strains of the tree for replanting.

- Unlike many trees, chestnuts germinate rapidly in spring, at around 68°F (20°C).
- Seeds sown in the fall require a thick mulch to protect them from freezing in winter.
- Due to the presence of the spores of chestnut blight, there is little point in trying to grow the trees in their historic native range. However, they can thrive in zones with lower summer humidity, such as the Pacific Northwest.

Chestnuts played an important role as a source of carbohydrate for the Native American communities of the Appalachians and other areas where the tree grew. The nuts were also very popular in the nineteenth century as a street food in the United States, being roasted on open fires.

A strong, straight-grained timber, American chestnut is also full of tannins, so rots only slowly in contact with the ground; it is thus popular for fencing. Today, reclaimed timber is still available; some is used for furniture.

Early European settlers in the United States made a coffee substitute from roasted dried chestnuts. Caffeine-free chestnut coffee substitute is still available today.

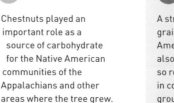

Up to 80 ft. (25 m)

Up to 100 ft. (30 m)

Castanea sativa

Sweet Chestnut

A large-growing and long-lived deciduous tree of
southern Europe, sweet chestnut has had major
economic importance for humanity for millennia.
It was distributed widely by the Romans, including to
much of northwest Europe, which is beyond the area at
which it can reliably reproduce. Growing naturally to
115 feet (35 m), the vast majority of Europe's chestnuts
were cultivated as coppice, which exploits the tendency
of the tree to send up a mass of straight shoots after
being felled. A huge decline in coppicing has resulted in
very dense, unmanaged woodland growth.

The toothed leaves, cream flower clusters, and densely
spined fruit are all characteristics that make the tree very
distinctive. Even the bark, with its pattern of spiral ridges,
cannot be mistaken for that of any other species.

Chestnuts were a
staple source of
dietary starch for
the people of many
European regions until
the introduction of the
potato from America in
the sixteenth century
signaled its slow but
irreversible decline—
hastened by the arrival
of cheaper grains.

Stems from coppiced
chestnut trees are cut
to make stakes for
fencing. The stakes
are linked top and
bottom with wires so that
the fence can be easily rolled
up and used as many times
as is necessary.

Up to 100 ft. (30 m)

Up to 115 ft. (35 m)

Chestnuts are the basis of
traditional treats, called
marrons glacés in France,
consisting of the nuts
candied in sugar. The nuts
may be eaten on their own
or used in cooking, often as
an ingredient in cakes and
tortes, chestnut cream, and
ice cream.

• Choose named varieties if nut
  production is of paramount
  importance. The nuts germinate
  easily and seedling growth presents
  no particular problems.
• Young trees should be planted
  23 feet (7 m) apart in full sunlight.
• The trees thrive best in a deep,
  fertile soil.
• Degenerated coppice may be
  revived simply by felling trunks.
  Thin but usable stems will be
  available in a matter of years.

Casuarina equisetifolia ▲

Beach She-oak, Australian Pine

Casuarina equisetifolia is one of seventeen species of She-oak, a distinctive group of trees that thrive in difficult environments. Growing to a maximum height of 100 feet (30 m), its natural range extends from Southeast Asia southward into northern and eastern Australia, and also includes some islands in the Pacific Ocean. Beach she-oak has been introduced into a great many warm-climate countries as an agroforestry tree, but in some places it has become aggressively invasive, crowding out competing native trees.

She-oaks are distinctive in having minute leaves on plentiful twigs. These make them look a little like conifers, an impression that is reinforced by the conelike appearance of the fruit. The ability of the trees to grow on soils too infertile for other species is helped by their having a symbiotic relationship with a fungus that grows on the roots. In exchange for carbohydrates provided by the tree, the fungus converts atmospheric nitrogen into nitrates, which the plant can then use.

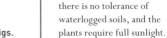

- The small seeds germinate quickly and easily.
- Growth is rapid, and the trees thrive in a wide range of conditions, although there is no tolerance of waterlogged soils, and the plants require full sunlight.
- The trees can be planted close together, and the first usable ones may be extracted from the rest in as little as five years.

Beach she-oak is a source of cheap timber of reasonably good quality that is used for basic construction projects and farm purposes such as shingles and fencing. It can also be used for pulp. By-products from milling, such as sawdust, may be added to soil to reduce its alkalinity.

As a fast-growing tree, this species is ideal for planting for erosion control and creating windbreaks. However, this must be set against the danger of it seeding and becoming invasive—it may not be planted on agricultural land for this reason.

Up to 50 ft. (15 m)

Producing high temperatures, beach she-oak makes good firewood. Increasingly, it is likely to be chipped for use in electricity generation. As biomass it clearly has potential for serving both local and distant markets.

Up to 100 ft. (30 m)

Catalpa bignonioides

Catalpa

Native to central and eastern North America, this rapidly growing deciduous tree is capable of reaching heights of around 65 feet (20 m) and has showy, white, tubular flowers and distinctive, broad foliage. The leaves are luxuriant and large, measuring perhaps 12 by 8 inches (30 x 20 cm); if trees are pollarded, as they often are in yards, the leaves grow even larger, creating a lush, tropical effect. The flowers give way to long beanlike seedpods that are almost unique in the woody flora of North America. These pods are the origin of a common name for the tree sometimes used in the United Kingdom: the Indian bean tree. The species first arrived in Britain in the eighteenth century and it remains a popular ornamental there—demonstrably more so than it does in its natural habitat of North America.

Catalpa wood is not used much as timber these days because the tree does not lend itself to long, straight cuts. In pioneer days, however, young trees were a common source of fence posts because the wood does not rot easily in the ground.

Catalpa leaves have an unusual use: as a food source for the caterpillars of the catalpa sphinx moth, which are renowned as a fishing bait throughout the American South. Trees are actually grown by farmers to feed the caterpillars.

A tea made from catalpa bark is reputed to have a mildly sedative effect, and has been used to treat troublesome coughs and malaria; it offers an antidote to snake bite, too.

- Catalpa seed germinates rapidly and prolifically.
- This is a pioneer species, so young trees planted in full sunlight will grow rapidly in any reasonably fertile soil.
- The species is hardy to around -4°F(-20°C) but has to be sheltered from strong winds.
- The tree is allelopathic, meaning that it may impede or prevent the growth of other plants around it.
- Catalpa trees cannot be shaped by pruning, but they respond vigorously to pollarding.
- The tree can tolerate both dry conditions and flooding.

Up to 40 ft. (12 m)

Up to 65 ft. (20 m)

Cedrela odorata

Spanish Cedar

A semi-evergreen tree that is widely distributed in South America, growing to a maximum of 165 feet (50 m), Spanish cedar is closely related to mahogany, with which it often grows in seasonally dry forest habitats. It is one of those trees that tends to be scattered through the forest and is never found in quantity. Although not well known outside the tropics, it is of considerable commercial importance. The timber is famed for its insect-repelling properties, but ironically commercial growing is limited by a number of insect pests, which make it uneconomic as a plantation tree; consequently it has been much reduced in the wild by felling. With large pinnate leaves, this attractive tree has been planted for shade in the Caribbean region.

- Fresh seed germinates rapidly and well.
- Young trees grow rapidly if conditions are suitable but have shallow root systems. The trees appear to grow most successfully on shallow limestone soils.
- Growing the tree in Africa has been particularly successful because fewer pests are present.

The heartwood contains a resin that acts as a deterrent to insects yet seems pleasantly scented to humans. Thanks to this useful combination of qualities, the wood has a long history in the manufacture of chests intended for storing clothes in tropical climates. It is also used in guitar and cigar box making.

Essential oil derived from the bark is used in folk medicine traditions to treat malaria and diabetes, wounds, fever, bronchitis, indigestion, and other gastrointestinal ailments. It is also used in medications for expelling parasites from the gut.

Up to 165 ft. (50 m)

Up to 165 ft. (50 m)

Being termite-resistant, Spanish cedar wood is often used in the tropics for construction, even though it is a relatively weak wood for that purpose, due to its low density. It may also be used for veneers and as a component of plywoods.

Despite its name, Spanish cedar is not only a hardwood (and therefore not a cedar at all), but it is also not Spanish.

"SPANISH CEDAR"
MCILVAIN WEBSITE

Deodar

This conifer grows to 165 feet (50 m) in the foothills of the Himalayas, from Afghanistan across to western Nepal. It is the national tree of Pakistan. It has the layered foliage typical of cedars, but overall a distinctively conical shape. Unlike most conifers, the cones disintegrate when mature to release the winged seeds. The tree's name is derived from the Sanskrit *devadaru*, meaning "tree of the gods," which is also the origin of its name in north Indian languages—clearly an indication of the importance of the tree to the people of the region. Deodar forests were, and still are, places chosen by Hindu mystics for meditation. Since the nineteenth century the deodar has been a popular tree for landscape planting in Europe and in North America, with introductions being made from different parts of its range.

- Seed germinates easily if sown in fall, or given a month's stratification.
- Seedlings are prone to rotting, so should be potted on soon after germination; then grown on for a year before planting out.
- The trees thrive on acidic soils with good drainage in full sun.
- They are hardy to -4°F (-20°C), with variation between cultivars.

Oil from the bark of all cedars is effective as a natural insecticide and fungicide and can also be used against slugs and snails.

The timber is one of the better ones available in its habitat and historically the rot-resistant hardwood has been used for construction —most famously for the houseboats of Kashmir. However, it is not strong and its use is likely to stay local.

The oil that gives all parts of the tree their distinctive scent has been used in many kinds of products for its antiseptic qualities. A drinkable potion may be distilled from the wood for the treatment of lung and urinary disorders, and fevers.

Up to 50 ft. (15 m)

Up to 165 ft. (50 m)

Ceiba pentandra

Kapok

This large semi-evergreen tree tends to have a fine straight trunk, covered by distinct broad-based spines; buttresses develop as it gets older. The leaves are palmate, and the flowers, with a mass of pink or white stamens, are usually borne during the dry season when the tree is leafless. These are followed by fruit that bursts open to produce seed attached to very light fibrous strands, which allow it to blow away on the wind. The fruits will also float for long distances if they land on water.

With a maximum height of 230 feet (70 m) it is not surprising that the tree was seen as connecting heaven, earth, and the underworld by the Mayan people of Mexico and Guatemala. The tree has particular significance for groups who are trying to revive Maya spirituality.

- Seeds germinate easily at temperatures of 68–77°F (20–25°C).
- Growth is rapid and successful on a wide variety of soils.
- Drought is tolerated, but not frost or prolonged cold.
- If grown in open conditions, a short, bushy tree may be expected; if grown in light shade with other trees, a taller, straighter trunk is more likely to develop.

The seeds are crushed to produce an edible oil that can be used for cooking, burned for lighting, and used as an ingredient in soap manufacture.

The fiber that envelopes kapok seeds is light and buoyant—eight times lighter than cotton and five times more buoyant than cork. Being also water-repellent, in the past it has had a wide variety of uses, including as a stuffing for mattresses, cushions, and pillows, and also life-jackets.

Unlike many stuffing materials used for soft toys, kapok does not tend to clump over time, and it also keeps its shape after washing.

It is clearly a sacred tree . . . a world tree upholding the world.

"THE WORLD TREE OF MAYA
RELIGION AND COSMOLOGY" ON
MAYA ARCHAEOLOGY WEBSITE

Up to 115 ft. (35 m)

Up to 230 ft. (70 m)

Celtis spp.

Southern Nettle Tree

The multiple species of the *Celtis* genus are all known as hackberries. Among them is the southern nettle tree, *C. australis*, which grows to 80 feet (25 m), although usually much less, and is found wild across central and southern Europe, Turkey, and in parts of North Africa. The very similar American hackberry (*C. occidentalis*) is found across the eastern half of North America.

With deciduous, nettlelike leaves and a roundish shape, this species is strangely undistinguished, with no showy flowers or fruit, or any particularly strong or distinctive characteristics; few people even notice it. However, the species has a long history in landscape planting because it is very tolerant of air pollution; as a result it was frequently planted in parks and urban areas in central European countries during the twentieth century.

- Seed needs three months of stratification or sulfuric acid treatment to facilitate germination.
- The tree will tolerate a wide variety of situations, including wet, dry, and poor-quality soils.
- In nature both *C. australis* and *C. occidentalis* tend to be found on drier soils.

In North America, wood from *C. occidentalis* was traditionally used to light the sacred fire that was used by the Apache tribe in ceremonies using the hallucinogenic peyote cactus.

The bark of *C. australis* has been processed to yield a yellow dye that is less toxic than many synthetic dyes.

Ripe fruits quickly attract birds, and are also edible by humans. It has been suggested that this is the fruit referred to as "lotus" by writers of ancient Greece, such as Herodotus.

Up to 65 ft. (20 m)

Up to 80 ft. (25 m)

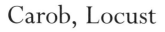

Carob, Locust

As a reliable source of animal fodder, carob is one of several tree species that were spread widely around the Mediterranean for about 4,000 years. Its glossy dark foliage, each leaf divided into five or six leaflets, remains a distinctive feature of many locations around the region. Growing to 50 feet (15 m) at most, it is a sturdy, solid-looking tree with deeply furrowed bark. It produces small flowers on catkinlike structures in fall; these are both wind- and insect-pollinated. The fruit is a podlike structure, 4 to 12 inches (10–30 cm) long with a woody appearance, that takes a year to ripen. The seeds are not edible (unlike those of most members of the pea family), but the pod is filled with a fibrous pulp which is distinctly sweet.

- Fresh seed germinates relatively quickly. Stored seed needs to be chipped or soaked, first in dilute sulfuric or hydrochloric acid and then in water, until it swells.
- The growth of young plants tends to be fairly slow.
- Transplanting should be done within the first two or three years of life.
- Fruit can be expected after five years.
- Most soils in full sun are suitable, and there is tolerance of light overnight frosts.

The seeds are remarkably uniform and were used in Classical times and in India as weights for precious materials. The "carats" now used to describe the quality of gold derive from the use of carob seeds.

Long since eaten in the Mediterranean, the sweet pulp of the carob makes a tasty alternative to chocolate. Like the date, it was a good food that was easily transported.

Carob is an ingredient of many desserts in the Middle East. The modern food industry mostly uses it as a sweetening and thickening agent.

Up to 65 ft. (20 m)

Up to 50 ft. (15 m)

Chamaecyparis obtusa

Hinoki, Japanese Cypress

Soaring to a height of around 115 feet (35 m) and with a trunk 3 feet (1 m) in diameter, the hinoki or Japanese cypress can be a magnificent tree. Japanese forests were originally full of these conifers and must have been magnificent, but virtually all have now been felled. They are being replaced by younger ones, this being a very important timber tree in Japan.

Like all cypresses, it has scalelike leaves, and here the cones are very small. A large number of garden cultivars (in excess of a hundred) exist, however, and these tend to have very different characteristics of foliage and growth pattern. Those with slow-growing or congested foliage forms are usually the result of propagation specifically from "witches' brooms," or growths that sometimes occur on conifer branches. On the other hand, the many forms with golden foliage are the result of natural mutations, which are then propagated through cuttings.

The wood is saturated with oils that preserve it from both rotting and insects. The oils have an attractive lemony scent that adds considerably to the value of the timber. The oil is extracted and used in aromatherapy and perfumery, and the wood is used for applications where its scent will be appreciated. Japanese hot-tub users, for example, regard the wood's scent, soft touch, and visual beauty as part of the experience.

- Seeds will germinate after stratification over a long period—for months or even years.
- Seedlings need dappled shade to begin with but full sun when established.
- Plants flourish in any reasonably fertile soil and there is a wide range of climatic tolerance.
- Hinoki takes pruning equally well for hedging, bonsai, and the traditional Japanese art of niwaki, or shaping trees into pleasing forms.

Hinoki is a rot-resistant building timber of superb quality that has a long history in Japan; the Ise shrine, the center for the Shinto faith, is built of hinoki, as are many other historic buildings in the country.

The pale wood is not only durable but strong, making it useful for a variety of purposes. It is particularly sought after for making traditional Japanese boxes because the scent is absorbed by their contents.

Up to 100 ft. (30 m)

Up to 115 ft. (35 m)

Chrysophyllum cainito

Star Apple

The star apple tree, native to the islands of the Caribbean, grows to around 100 feet (30 m) and produces a spherical fruit with a diameter when ripe of 2 to 4 inches (5–10 cm). It is one of many minor tropical fruit that potentially have a more important commercial future. A long presence in Central and South America, and in the Philippines, indicates its spread through the Spanish empire. More recently, it has become popular in Southeast Asia. The fruit, which may be green, brown, or yellow, produces a latex from the inedible rind, and this is the source of some of the tree's popular names, which often allude to milk. The latex is actually bitter in flavor, and it must not be allowed to come into contact with the hands when the fruit is prepared.

- Fresh seeds germinate quickly to produce trees that fruit in five to ten years.
- Young trees grow relatively fast, and adapt well to any well-drained soil.
- There is no frost-resistance.
- As yet there has been very little varietal selection.

Folk medicine traditions generally use preparations of the bark or the unripe fruit for treating intestinal problems. The mucilaginous or gelatinous ripe fruit is a soothing remedy for laryngitis.

Star apple timber is not commercially exploited, but the growth in popularity of this fruit will increase the availability of timber as plantations are established. Potentially the wood is valuable, because it is hard, heavy, not difficult to work, and has attractive pink or purple tones that may find a market in flooring and furniture making.

The fruit is sweet and tastes best if served chilled. In the West Indies it is often combined with other fruit in chilled desserts; a Jamaican combination with sour orange is called "matrimony."

Star apple is one of the very low calorie exotic fruits.

HEALTH BENEFITS
WEBSITE

Up to 100 ft. (30 m)

Up to 100 ft. (30 m)

Quinine

This small-to medium-sized evergreen tree, never more than 50 feet (15 m) tall, has, since the earliest days of Western exploration of the Americas, been the most important source of quinine, the most effective natural treatment for malaria. Its medicinal properties had been discovered a long time before the arrival of the Europeans by the healers of the Quechua people of its homeland, the warm, humid forests of the lower slopes of the Andes. In particular, the alkaloids it contains in its bark are very effective at reducing fever.

The tree has pointed ovate leaves and clusters of tubular white or pinkish flowers, the shape being very typical of the Rubiaceae family. There are many other *Cinchona* species, and because they hybridize readily their naming is confused; current research aims at resolving species definitions and origins through DNA analysis.

- Seed is the main method of propagation. Grow on plants in light shade in a climate where temperatures do not fall below 46°F (8°C) or above 79°F (26°C).
- An annual rainfall of at least 100 inches (250 cm) is regarded as most favorable.
- Waterlogging and drought are both harmful.

Quinine as a drug is the most effective of the many herbal treatments that have been used over the centuries to treat both the symptoms and cause of malaria. Consequently quinine production has motivated much endeavor, and destruction of trees, with major deforestation and social exploitation in its homeland. In World War II, quinine plantations in Southeast Asia were considered of vital importance and were the subject of many battles.

Derived from a malaria prophylactic, the bitter taste of quinine is now regarded as essential to a well-made gin and tonic.

Up to 35 ft. (10 m)

Up to 50 ft. (15 m)

Camphor

Growing to 100 feet (30 m) and potentially living for more than a thousand years, the camphor tree is an impressive species. Some of the best specimens are to be seen in the grounds of Buddhist and other temples in China, where they have been protected. The evergreen leaves are oval, with a very glossy green and a powerful smell when crushed. The species grows in southern China, Japan, and Vietnam, but it has been introduced widely across warm-climate regions; in some places, such as Australia, it has become dangerously invasive.

One reason behind the invasive qualities of the tree is that its leaf litter suppresses surrounding plants; another is that the berries are eaten by birds, which distribute the seeds widely. Ironically, the tree is now very rare in the wild in China.

- Seeds lose viability quickly; they germinate over months at 68–77°F (20–25°C).
- Young plants need to be transferred early.
- Warm temperatures are needed, but there is tolerance of drought, wind, and light overnight frost.
- Most well-drained soils are suitable.

The camphor in the wood deters or even kills pests such as moths. It has long been used to keep clothes, documents, and other items safe. It was favored for making sea chests, which protected against boring insects.

Camphor is a waxy material that is extracted from all parts of the tree, but primarily from the leaves, through distillation. It is used as an ingredient in aromatic balms for the treatment of colds or aching limbs. It may also be used on the skin to relieve pain and itching.

Up to 65 ft. (20 m)

Up to 100 ft. (30 m)

The wood is fine-grained and dense, so is suitable for detailed carving. The camphor content helps prevent decay. Some of the oldest carvings in Japan are made from the wood.

Cinnamon

Growing to 50 feet (15 m), this relatively modest-sized evergreen tree dominates the world spice trade. A native of Sri Lanka (whose old name of "Ceylon" lives on in the specific name *zeylanicum*), this is the source of much of the world's "true" cinnamon. Other species of *Cinnamomum* also contribute to the cinnamon market, and there is relatively little to choose between them in spicing cakes, curries, pancakes, and the rest of the vast array of food items that are flavored using cinnamon around the world. All take advantage of the sweet, warm flavor contained in what is actually tree bark.

Sri Lanka still produces the vast majority of the world's supply of the "true" spice, with China, India, and Indonesia producing the bulk of the "other" cinnamons. Much of their product is sold as rolls or "quills" of bark, with the Western market being catered for largely with ground bark.

Cinnamon bark rolled into quills is in many ways more convenient when cooking savory foods such as curries or rich dishes. The quills can be broken into lengths appropriate to the level of spicing required, and are then easily removed when serving.

Cinnamon is used as an ingredient in home cold remedies. Research indicates useful effects for treating diabetes, viral infections, and Alzheimer's disease.

- Seeds lose their viability quickly, and germination may take several months at 68–77°F (20–25°C).
- The tree flourishes at temperatures of 68–86°F (20–30°C), with an annual rainfall of 50 to100 inches (125–250 cm), although it can tolerate a dry season.
- The tree prefers light soils enriched with organic matter.
- For producing the maximum amount of cinnamon bark, trees are often coppiced.
- The tree can be grown in bonsai form.

Up to 50 ft. (15 m)

Up to 50 ft. (15 m)

Specialists are debating the maximum levels of cinnamon that should be used in foods. There is a possible harmful effect of one compound, coumarin, for a small number of susceptible people.

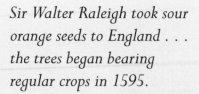

Sir Walter Raleigh took sour orange seeds to England . . . the trees began bearing regular crops in 1595.

JULIA F. MORTON,
FRUITS OF WARM CLIMATES (1987)

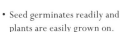

Bitter Orange, Seville Orange

Also known as the Seville or sour orange, this tree produces fruit that is bitter and inedible, but which when cooked with sugar produces the familiar marmalade of the breakfast table. A hybrid between *Citrus maxima* (pomelo) and *C. reticulata* (mandarin), the tree is thought to have originated in either northern India or southern China, or possibly both, as the species comprises at least two distinct genetic lineages. Growing to a maximum of 35 feet (10 m), it is an evergreen, with dark foliage and sweetly scented white flowers. In Mediterranean-climate regions, orange groves are often an important part of the landscape.

Most cultivars produce fruit that is too bitter to eat, but this is the cultivar from which the vast majority of orange-flavored products are made. The reason is that the sweet oranges that are eaten fresh lose their flavor if cooked. Consequently this is a plant of major economic importance. Like all citrus, it is susceptible to a number of pests and diseases that have a considerable impact on its productivity; some of these make it impossible to grow the tree in an otherwise favorable region. In the case of pests, biological controls are being actively pursued by researchers as an alternative to pesticides.

- Seed germinates readily and plants are easily grown on.
- Seedlings need sun and fertile, well-irrigated soil, but will only survive light overnight frosts.
- *Citrus* X *aurantium* is commonly grown in pots and protected over the winter in colder climates; attention needs to be given to the nutrition of potted trees.
- Great care must be exercised to avoid disturbing the roots when re-potting a tree into a larger container.

A range of alcoholic liqueurs, generally aimed at the luxury end of the market, are prepared from bitter oranges; many of these have originated from bitter orange varieties in the Caribbean, such as the laraha in Curaçao. The Caribbean region also offers a range of fresh cocktails that are prepared with the juice, typically using added rum and other local flavorings.

Citrus fruit skins are the source of D-limonene, an organic solvent that provides a relatively safe alternative to petrochemical ones in situations where human contact is likely. D-limonene is often used as a stain remover for oil-based stains, for removing glues and sticky substances from surfaces, and for removing grease and oil from the skin.

Up to 23 ft. (7 m)

Up to 35 ft. (10 m)

In parts of the Middle East the juice of sour oranges is used in stews, pickles, or in marinades, or just dribbled over food. The fruit are eaten in Iran and Mexico, and the dried peel is sometimes used as a seasoning.

One of the world's best-known breakfast products, marmalade is a conserve that brings together an ideal balance of acidity, bitterness, and sweetness. Many different flavorings have been added to widen its appeal. Seville oranges tend to be the main ingredient.

Kumquat

Kumquats are slow-growing miniature citrus trees that reach a maximum height of 15 feet (4.5m). Wild populations are still to be found in the evergreen hill forests of south-central China; however, the tree has been in cultivation for at least a thousand years in China, and for several centuries in Japan. The tree is commonly used as an ornamental, and cold-hardy varieties are popular in northern latitudes as yard plants, or in pots that may be moved inside to protect them from storms.

There are many varieties of *Citrus japonica*, but in the past a number of them were considered as belonging to another genus, *Fortunella*, named after Robert Fortune (1812–80), the British plant hunter who introduced the kumquat to the West in 1846. All of the *Fortunella* species have since been identified as genetic variations of *C. japonica*, and are therefore true *Citrus* plants.

Despite this scientific confirmation, the old name of *Fortunella* very much remains in use as a means of distinguishing different kumquat varieties. Today, the four varieties most cultivated for culinary use are the oval Nagami (*F. margarita*), the round Marumi (*F. japonica*), the large Meiwa (*F. crassifolia*), and the little Hong Kong Wild (*F. hindsii*).

- Kumquats are relatively easy to grow, but different varieties vary in cold hardiness and the degree to which they are thorny.
- The small size of the plants lends them to pot culture, which means that they are easily taken inside during cold winter weather, although the light levels inside have to be good for them to thrive.
- "Marumi" is the most cold hardy. Its fruit is extremely sweet and pleasant.

Kumquat pulp is diuretic and detoxifying, which makes it good in diets intended to cleanse the digestive tract, to reduce edema (fluid retention), or to treat gout or rheumatism. Kumquats also contain flavonoids that improve blood flow by conditioning the blood vessels. They benefit cases of hypertension, diabetes, high cholesterol, and heart disease.

Kumquat fruit can be eaten whole to enjoy a unique combination of sweetness and sourness, although they are not particularly juicy. The fruits can be used in a variety of dessert recipes or in preserved form, sometimes with honey, which is how they often feature in East Asian cuisine. Their sourness makes them delicious with bitter green salad leaves, such as endive, with kiwi, or as marmalade. Kumquats also make an unusual ingredient for smoothies.

Up to 13 ft. (4 m)

Up to 15 ft. (4.5 m)

As a New Year festive tradition in China, people exchange small bunches of kumquat leaves and fruits as a means of wishing one another gold and good fortune. Simply eating kumquat fruit is also a means of bringing prosperity.

Kumquats or preparations made from them are rich in vitamin C. In the traditional medicine of East Asia they are primarily used to treat colds, coughs, sore throats, and chest disorders.

It is the only citrus fruit that can be eaten "skin and all." The peel . . . can be eaten separately.

KUMQUAT GROWERS WEBSITE

Citrus × *limon*

Lemon

Very similar in appearance to orange trees, those of the lemon grow to around 23 feet (7 m), and flourish in a similar climatic range. The origin of the species is still a mystery; it is thought to have first appeared in northern India as a hybrid between the citron (*Citrus medica,* see opposite) and the bitter orange (*C.* × *aurantium,* see p. 68). Traveling west, it was brought to Europe by the Arabs, via Spain. The first mention of the fruit is in tenth-century Arab writings. With myriad uses, the lemon has diversified into a wide range of cultivars; these are an important crop in many regions, with China, India, and Mexico leading global production.

As with other citrus, the fruit takes a long time to ripen, so flowers and ripe fruit are often seen together on the tree. The color typically thought of as "lemon yellow" only develops in cool winter-climate environments—fruit grown in the tropics stays green.

Oil extracted from the peel of lemons makes a very effective and safe pesticide; its action is effectively to suffocate insects and mites, because two properties of citrus fruits, d-limonene and linalool, act as natural nerve toxins in insects.

Used as a flavoring of classic North African cuisine, pickled lemons are prepared by stuffing salt into cut fruit and packing them into jars. Their unique flavor develops through anaerobic fermentation during the following few months.

Up to 15 ft. (4.5 m)

Up to 23 ft. (7 m)

The acidity of lemons makes them a good cleaning agent; for example, they remove tarnishing from metal, soap scum and limescale from bathrooms, and dirt from tile grouting. Their smell is valuable, too, because they create a clean and fresh atmosphere.

- Cultivation is very similar to the orange, although there is considerably greater variation in key issues for gardeners, such as tree size and hardiness.
- "Meyer Lemon" is a small-growing and relatively frost-hardy variety ideal for growing indoors in cooler climates. It may also be risked outdoors if a sheltered spot can be found.
- "La Valette" is a particularly good small variety.

Citron

The citron was the first citrus fruit to make its way from Asia toward Europe. Seeds that are at least six thousand years old have been found in archaeological sites in Mesopotamia. The Romans grew the tree, as did the ancient Israelites; Jewish communities, in particular, may have distributed it around the Mediterranean.

The tree is vigorous, thorny, and straggly, and grows to 15 feet (5 m). Its origin is unclear, but its ancestor, probably from northern India, is possibly now extinct. Its large fruit is rarely eaten directly, but its aromatic rind and plentiful pith held a fascination for people in an era when other exotic fruits were unknown. Many different varieties are known; particularly bizarre is one known as the "Buddha's hand," which has fruit divided into many fingerlike appendages.

- The citron is a slightly less tolerant of frost and drought than some other citrus fruit, but is otherwise very similar in cultivation.
- It can be grown from pips if these can be obtained, or plant leafy cuttings from two- to four-year-old branches deeply in the soil.

The citron plays an important role in the Jewish Feast of the Tabernacles, and has become a symbol of that faith. Communities north of the region where it was grown always made great efforts to obtain it. The growing of ritual fruit was overseen by a rabbi, with strict rules governing their shape and quality.

Jams, pickles, and other preserves are made from the pith in the Middle East, Pakistan, and northern India. The "Buddha's hand" variety is steamed and eaten in East Asia.

Up to 12 ft. (3.5 m)

Up to 15 ft. (5 m)

Traditional medical systems made use of the citron for many ailments, including nausea and chest and intestinal problems.

> *Kerala derived her name from Kera—the coconut tree, called land of Kera and true to her name she owes her existence to coconut palms.*
>
> ON PALAKKAD COCONUT PRODUCERS COMPANY LIMITED WEBSITE

Cocos nucifera

Coconut

The coconut is a palm, growing to around 100 feet (30 m), that is thought to have originated in the Pacific region but that has been spread around the globe—for this is an extraordinarily useful plant. Traditional peoples who live in coconut-growing regions have found a use for practically every part of the plant. The trees are also quite extraordinarily productive; the flesh of the coconuts, which each weigh about 3 pounds (1.4 kg), has always been an important food source, but in recent times many other uses have been found both for the flesh and the oil it contains.

Coconuts are one of those crops that appear to just grow themselves, without assistance from humankind, which indeed they do; however, there is a great deal that can be done to improve their productivity. Much research is currently being done on breeding and best cultivation practices; in many countries this is only the start, because what is discovered then has to be disseminated to smallholders and farmers.

- Coconuts placed in contact with warm, moist soil take several months to germinate, after which growth is continuous and rapid in the appropriate conditions.
- They are completely tropical plants, not thriving anywhere that gets cold. A mean annual temperature of 81°F (27°C) is optimal.
- They also need plentiful water, with an annual rainfall of over 39 in. (100 cm).
- Any reasonably well-drained soil is suitable for the coconut.

Like all palm woods, coconut does not have a grain, but it does have a very distinct patterning that makes it a good material for relatively small items, either functional or decorative. The wood comes in different densities—the highest being suitable for load-bearing and therefore used in construction or for flooring, and lower densities used for paneling and cladding. The wood is also used to make distinctive furniture.

While nowhere in the world is coconut regarded as a staple food, the flesh is a major source of both calories and flavoring in many cuisines. For example, in southern India the flavor is prominent in many stews and relishes, but there are also coconut-centered dishes like vattayapam, a cake made of rice and ground coconut prepared through a process of first fermenting and then steaming.

One of the major drivers to increased production is the growing use of coconut oil in cosmetics in countries far outside the "coconut belt." The oil is traditionally used neat on the hair, but contemporary processes incorporate it as an ingredient into a wide range of products for skin and hair care.

Coconut leaves can be woven into many domestic items, such as hats, containers for food, disposable crockery, and floor coverings, while the presence of strong fibers means that they can also be used for making stiff broom and brush heads.

Up to 35 ft. (10 m)

Up to 100 ft. (30 m)

Commiphora myrrha

Myrrh

Myrrh is familiar to most people as one of the gifts brought by the Three Kings for Jesus in Bethlehem. It served as a reminder of his mortality, for the gum exuded by the tree has long had a role in embalming the deceased.

Myrrh is an aromatic yellowish gum that exudes from wounds in the bark of a small, deciduous, and very spiny tree found in the Arabian peninsula and the Horn of Africa. It grows to only 10 feet (3 m). The tree has a swollen trunk that enables it to store water and so survive drought. The gum is collected from wild bushes that are repeatedly cut in order to stimulate production of the material. Myrrh has been traded out of its homeland for centuries, and today several species of *Commiphora* are tapped on a large scale for international trade.

- Commercially obtainable seed generally has a very low rate of viability.
- The plant may be propagated from cuttings.
- Over-watering or poor drainage are anathema, and the plant has a tolerance of poor, stony, dry soils and long periods of drought.
- The leaves and buds have a strong aroma, as well as the gum.

Several traditional medical systems, including those of India and China, have found uses for myrrh. Research indicates that it can be used to help treat diabetes, and in laboratory tests it has been shown to slow tumor growth in cancers. It is also used against coughs, asthma, and congestion.

Long traded as an ingredient of incense, myrrh has also been used in perfumery, where its smoky balsamic tones have great value. Some prefer the sweeter scent of frankincense and find that of myrrh harsh and medicinal.

Up to 10 ft. (3 m)

Up to 10 ft. (3 m)

In the early years of Christianity, incense was expressly forbidden.

J. COHEN, "A WISE MAN'S CURE"

The gum has had a long association with religious ritual. In the Eastern Orthodox Church, it is used to scent chrysm, an oil that is used in administering the sacraments. Myrrh is often mixed with frankincense for ritual purposes.

Cornelian Cherry

A small deciduous tree or large shrub that grows to 40 feet (12 m), the Cornelian cherry is found throughout southern Europe, the Middle East, and the Caucasus. Small clusters of pale yellow flowers appear on bare branches in late winter or early spring and are followed by the formation of an almost rectangular red berries, ¾ inches (2 cm) long, which contain a single seed. Cultivars selected in the former USSR can have fruit twice this size. The tree is popular in ornamental horticulture in northern Europe for its early flowering, although in this region its fruiting is limited to hot summers. The leaves turn a dark purple in the autumn months; variegated and golden foliage forms have also been selected.

- Seed should be rigorously separated from the fruit pulp, and sown as fresh as possible; otherwise, give them four months of stratification.
- Stored seed may be very slow to germinate.
- Young plants need some winter protection for the first year. They are easily grown in any reasonably fertile soil, but fruit best in a warm, sheltered location.
- The tree is hardy to -4°F (-20°C).

Flesh wounds and ulcers can be successfully treated with preparations of the fruit and an oil derived from the seeds. Traditional medical systems use flower and fruit preparations in treating a number of infectious diseases.

The berries are edible when ripe, with a pleasant acidic flavor. In eastern Europe and the Middle East these berries are used to make jam and cooling summer drinks, and in various ways to flavor spirits and liqueurs.

Up to 24 ft. (7.5 m)

The wood is one of the hardest to be found in Europe, and has been favored for the manufacture of machine parts, tool handles, and particularly weaponry.

Up to 40 ft. (12 m)

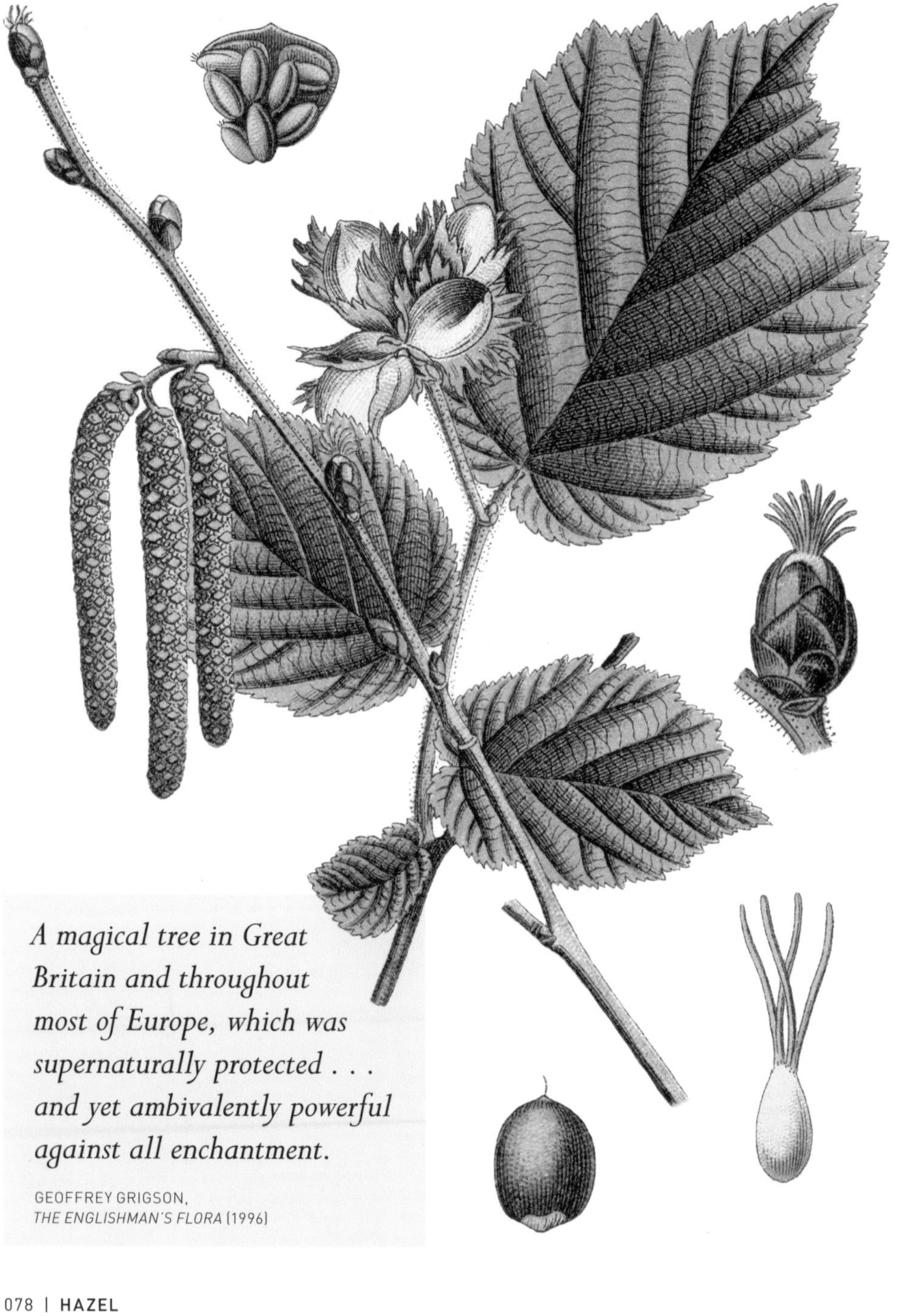

*A magical tree in Great
Britain and throughout
most of Europe, which was
supernaturally protected . . .
and yet ambivalently powerful
against all enchantment.*

GEOFFREY GRIGSON,
THE ENGLISHMAN'S FLORA (1996)

Corylus avellana

Hazel

Regarded by many as a deciduous shrub because it continuously regenerates by producing new branches at the base, hazel can reach tree proportions, as high as 50 feet (15m); however, half of that is more usual. A species of major ecological importance, it is primarily an understory plant of mature woodland on fertile soils across Europe, as far as the Urals and down to northern Iran. It has broadly oval, toothed leaves, prominent, wind-pollinated, catkin-type flowers in spring, followed by small nuts that are an important resource for woodland wildlife. Rodents such as squirrels are well known for stashing nuts for eating in the cold months, and these, if not recovered, go on to germinate—a good way for the tree to regenerate.

Although the nuts offer high-value protein, the main traditional economic value of hazel lies in its timber, produced by the coppice system, which makes use of the plant's habit of continually producing basal shoots. These are extremely straight for their first few years, while older ones still make good, if not such regular, lengths of pole. Coppicing traditionally promoted a regular cycle of growth and involved cutting the tree back to its base every ten or twenty years.

- Hazelnuts need chilling for four months at around 39°F (4°C) as a pre-treatment before being sown.
- The plants can also be propagated by grafting (which would be used for ornamental foliage varieties), the pulling off of rooted suckers, and layering.
- The trees flourish in fertile moist soils, and are tolerant of sunlight or light shade at northern latitudes.
- Trees bearing nuts are highly attractive to squirrels.

Hazel rods cut in the first few years after either coppicing or natural regeneration are about as straight as any natural product can be. They have traditionally been used to make fencing, in the making of crude furniture, and in wattle-and-daub construction, whereby clay and mortar is applied to a lattice of hazel branches to build a wall.

Hazel twigs, cut during the winter, have long been used as a support for growing garden peas because they usefully offer multiple small twigs in a flat plane; crucially, they do not root and so start competing with the peas for soil nutrients, as would supports made from willow.

Forked hazel twigs have been the instrument of choice for dowsing, a traditional practice that enables supposedly sensitive individuals to detect underground water.

The high protein content of hazelnuts helps to account for their rich flavoring, which is made use of in European and Middle Eastern cuisines, overwhelmingly for sweet dishes or pastries. Hazelnuts can be used to make praline, a popular filling for chocolates or pastries in central Europe.

Up to 46 ft. (14 m)

Up to 50 ft. (15 m)

Crataegus monogyna ⚠

Hawthorn

Often regarded as a hedging shrub, hawthorn is in fact a long-lived deciduous tree that grows as high as 50 feet (15 m). Found across Europe as far as Ukraine, this species is a key component of the scrub that takes over neglected pasture, particularly in the north of its range. A mass of white flowers in late spring can be a dramatic sight in places where it is common; these turn to bright red fruits that ripen by late summer and are a major food resource for migratory birds.

The flowers have an odd smell, and it has been established that one of the chemicals released is the same as one of those responsible for the odor of putrefying flesh. Historical accounts associate the scent with that of the Black Death in medieval Europe. The scent may possibly account for a range of beliefs that associated the tree with death and the underworld. Indeed, hawthorn has long had a close relationship with humanity, and in some cultures, especially Celtic ones, an array of magical beliefs are associated with it. In the British Isles it is the main component of the extensive system of hedges planted in the eighteenth and nineteenth century to demarcate field and property divisions.

- Seed should be sown in fall, to chill over winter—a few seedlings will emerge in spring, but most will not until another year has passed.
- Young trees grow rapidly (accounting for another common name for them – quickthorn), and they are remarkably tolerant of a wide range of conditions, including windy sites, but will not flourish on very wet ground.
- Hawthorn hedges can be cut and shaped easily, but due to their speed of growth they must be trimmed frequently or they will rapidly develop into full-sized trees.

Fresh and green young hawthorn leaves in spring make a subtle-tasting but pleasant addition to salads. *Crataegus* juice is manufactured into a range of fruit drinks and soft drinks, and into a vinegar product. A teaspoon of dried berries may be made into a refreshing cup of tea simply by adding boiling water.

Hawthorn grows fast and responds well to clipping, so it makes a good hedge that can be kept relatively formal and tidy. Alternatively the hedge may be maintained through the traditional practice of "laying," where the trunks are split and then laid down along either side of the hedge, where they resprout. This technique makes for a strong and effective stockproof fence, reinforced by the hedge's many stiff thorns.

Up to 20 ft. (6 m)

Up to 50 ft. (15 m)

The density of hawthorn wood makes for the hottest fire of any north European tree species, and its charcoal is so high in carbon that it has even been used for small-scale iron making.

The berries are used for heart problems, and there is evidence that hawthorn benefits circulatory problems, particularly if used alongside drug-based treatments.

Thorn is the most common tree-name
constituent of English place names.
There are numerous Thornhills and
Thorntons as well as Thornbury,
Thorncombe, Thornham, and plain
old Thorns . . .

ARCHIE MILES, *SILVA, THE TREE IN BRITAIN* (1999)

Chinese Hawthorn

There are around two hundred species of crataegus or hawthorn distributed across the northern hemisphere. Many have been used as sources of herbal medicine or food by early hunter-gatherer peoples. *Crataegus pinnatifida* has one of the largest fruits and illustrates the potential of the genus.

Growing to a maximum height of 23 feet (7 m), the tree has the clusters of white flowers typical of the genus, followed by exceptionally large fruit, up to 1 inch (3 cm) across. The deciduous leaves are partially divided and very much larger than those of most other hawthorns. The tree is native to northern China and Korea and is found naturally on rocky slopes. The level to which the tree has the thorns typical of the genus varies—some specimens have almost none. While the species has considerable ornamental value, it is rarely seen in cultivation, perhaps because of its prominent role in agricultural hedging.

- Seed germination is slow, and may take more than a year.
- It is possible that seed collected "green," or before it is fully ripe, may germinate more quickly.
- Young plants should be planted out in their second year and no later.
- The species has a reputation for being easy to grow and very tolerant of soil type. It also tolerates both wet and drought.
- In the West, grafted plants can bear fruit at three years.

The large fruit has an acidic flavor and mealy texture. It is generally used for making preserves, such as jam, and in flavoring alcoholic drinks. In China its best-known use is in the making of *tanghulu*, where the fruits are threaded onto sticks and dipped in boiling sugar water to make a candied snack sold from street stalls in areas where the fruit is found.

As with other *Crataegus* species, there is evidence that the fruit has a role in managing certain cardiac problems, as well as hypertension. Traditionally, the fruit has been used as a digestive remedy.

Up to 20 ft. (6 m)

Up to 23 ft. (7 m)

Hawthorn was first mentioned as a drug in the Tang-Ben-Cao, a Chinese herbal [of 659 CE].

STEVEN FOSTER, "HAWTHORN," ON HERBALIST STEVEN FOSTER'S WEBSITE

Italian Cypress

One of the world's most instantly recognizable trees, the Italian cypress is a conifer that grows not just around the Mediterranean but also eastward into the Middle East, and as a wild species it can be found as far east as Iran. The familiar pencil shape of this tree is not strictly speaking a "natural" one, rather the result of plant breeders selecting out narrow forms over the millennia it has been in cultivation. Growing to 115 feet (35 m), it can also live to a great age—a thousand years at least.

The foliage is dark green, while the branches sweep strongly upright in most of the forms in cultivation. This habit of the branches makes them very vulnerable to damage by settled snow, whose weight can force down and break the branches. This is the only reason that the tree is not widely grown in northern areas, because *Cupressus sempervirens* is as frost hardy as most other large trees native to the Mediterranean climate zone.

- Italian cypress is usually propagated by cuttings to ensure the passing on of the distinct characteristics of the cultivated parent.
- The tree can also be grown from seed, which needs to be cold-stratified for thirty days before sowing.
- Young trees grow relatively quickly and need little after care—just irrigation in dry spells.
- Established trees do better on drier sites and do not need feeding.

Aromatic oils in the wood deter insects, such as clothes moths, so the wood was historically used for making storage chests for clothes. Today, furniture made from cypress wood is relatively rare, although the wood is still available in the form of small balls for scattering among fabric items that require safeguarding.

The wood is highly scented and of good quality, but the tree's slow growth rate does not make much of it available. This gives the wood a certain curiosity value, leading to its use for small, high-value items, such as carvings.

Up to 30 ft. (9 m)

Up to 115 ft. (35 m)

The planting of these dark-leaved, dignified-looking trees is a tradition in cemeteries, in both the Christian and Muslim worlds. Being long-lived, it is a symbol of immortality for both faiths. However, it is equally popular in Mediterranean countries as an ornamental that brings elegance to a garden or landscape.

Quince

This small deciduous tree grows to 23 feet (7 m) in height and has a fruit very similar in shape to that of the related and more familiar pear. Native to eastern Europe and central Asia, it produces fruit most predictably in a continental climate. The history of its cultivation is very ancient; it was grown by the Greeks and the Romans, and quite possibly in China, too, although in East Asia the very similar *Pseudocydonia sinensis* tends to be grown instead.

Quince fruit is hard, and only ripens in warm-summer climates. It is certainly inferior to that of the pear, which has displaced it as a cultivated plant, especially since the eighteenth century. The tree is now rarely grown commercially, although its comparative rarity has made it attractive as a garden ornamental, producing a novelty fruit that owners often make into jam.

- Named varieties are propagated by grafting or by taking hardwood cuttings.
- Grown from seed, there is no guarantee of fruit quality.
- The trees need a fertile, deep loam, and a warm sheltered aspect to protect the flowers from nipping by early frost.
- Pruning is initially similar to that of apples, but little more is needed after four years.

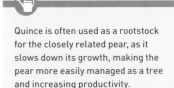

Even when not fully ripe, quinces are delicious if given a long, slow bake in the oven, preferably immersed in wine and flavored with spices. In many countries they are processed into "cheese," a solid and long-lasting sweet jelly.

Quince is often used as a rootstock for the closely related pear, as it slows down its growth, making the pear more easily managed as a tree and increasing productivity.

Eating an excess of quince can be dangerous because within the seeds are nitrogen-containing compounds that can be metabolized to cyanides.

Up to 35 ft. (10 m)

Up to 23 ft. (7 m)

Diospyros kaki

Japanese Persimmon

Reckoned to be one of the oldest trees to have been taken into cultivation, this species is found wild across a large area of eastern Asia. From the nineteenth century onward its cultivation was extended across the globe into many other warm-temperate or subtropical climate zones. Growing to 3 feet (10 m), it has broad deciduous leaves, but these have usually fallen by the time the fruit is ripe because it tends to hang, bright orange in color, on bare autumnal branches. The fruit are unpleasantly and famously astringent until they are fully ripe, when they develop a very soft, almost mushy texture.

This species has a particularly strong place in Chinese culture, being seen as a symbolically positive plant because of its longevity, productivity, and resistance to disease. Over time, many cultivars have been produced, including some with reduced astringency.

- Selected cultivars are usually grafted onto rootstocks; in the United States the related *Diospyros virginiana* is used, which has much greater cold resistance.
- Some varieties are reliably hardy to -4°F (-20°C), others to even lower, but only in a continental climate, however, because the trees need warm summers to ripen growth.
- Trees can begin to crop early, as young as three years old.

The fruit may be eaten fresh or used as an ingredient in a wide variety of dishes, including ice-cream, pies, desserts, and jam. The fruit can also be dried for use later in the year, as a snack or dessert, or, in Korea, as the basis for a punch.

Persimmon wood is hard (it is related to ebony) and has been used for making the heads of golf clubs.

In East Asia the fruit has been used to treat a wide variety of ailments, including stomach problems, high blood pressure, and coughs, but its efficacy has not been proved.

Up to 46 ft. (14 m)

Up to 33 ft. (10 m)

Dipteryx odorata

Tonka, Cumaru

This is an important tree of the rain forest canopy and is sometimes an emergent species. An evergreen growing to 100 feet (30 m), it is found in South America, chiefly in the Amazon basin. Research indicates that the trees may live for as long as a thousand years. The tree produces dark gray or black beans enclosed within green, pulpy fruit. The beans are up to 1 inch (2.5 cm) long, with a wrinkled surface texture, and have a very distinct fragrance owing to the presence of the chemical compound coumarin (the name comes from a French term for the tonka bean, *coumarou*). The fragrance is not dissimilar to that of vanilla, and tonka bean extract is sometimes used as a substitute for it. However, coumarin is toxic in large doses and there are strict legal controls governing its use. The anticoagulant medication Warfarin, which was initially created as a pesticide against rats and still used for that purpose, can be made from it.

- Little is known about the cultivation of the tonka tree and most specimens grow in the wild.
- Fresh beans may be expected to germinate within weeks.
- Further growth will depend on providing the continuous warm and humid conditions of its natural habitat.

Tonka timber, often available as Brazilian teak, is durable, hard, and dense. It has a reddish-brown color and in sunlight fades only slowly to silver-gray. It is used for construction and boat building purposes, including flooring and decorative moldings.

Tonka beans . . . are said to lighten one's mood and be emotionally balancing.

ON SCENTS OF EARTH
WEBSITE

Up to 70 ft. (21 m)

Up to 100 ft. (30 m)

In South American folk medicine, the beans and bark are used to treat fevers and a variety of other complaints, including mouth ulcers, earache, and sore throats.

Used in the past as food additives, tonka bean extracts have had a particular role in flavoring tobacco and snuff, but they are now banned for this purpose in the United States. Tonka is still used in perfume making, where its possible toxicity is not of concern.

Drimys, Canelo

An evergreen tree with lance-shaped green leaves from southern Chile and Argentina, drimys grows to 65 feet (20 m) and is a common component of a temperate rain forest community that is a distinctive feature of the region. Bunches of white flowers, which incorporate a surprisingly large number of petals, decorate the tree in early summer. The petal numbers are an indication that it is a very primitive species of flowering plant.

Flourishing in mild, moist climates, this is an ornamental tree renowned in western Europe, although it has never been extensively cultivated outside grand private yards. Growing at a modest rate, it keeps itself well covered with foliage down to the base of the plant and so tends to be seen as a shrub rather than a tree—the foliage rather hides the tree's rich red-brown bark.

- Seed takes over a year to germinate, at cool temperatures.
- Surviving very low temperatures and coastal exposure (it has even been grown as far north as the Faroe Islands), it can be damaged by cold but still grow back.
- It only develops to its full scale on moist, well-drained soils in high rainfall areas with cool summers and mild winters.

The entire plant has a spicy aroma and is used as a flavoring by the Mapuche people of Chile. The pungent, peppery ground bark has been promoted as a culinary spice.

The bark is one of the richest sources of dietary vitamin C, and was famously used by both Drake and Magellan to prevent scurvy among their crews on their pioneering global voyages. An unfortunate side effect can be severe diarrhea, which limits its usefulness.

Up to 25 ft. (8 m)

Up to 65 ft. (20 m)

The wood is heavy, with a pink or reddish color and a distinctive marbled patterning. It is used in the production of high-value items such as musical instruments.

Durian

Known as "the king of fruits," the yellow spiky durian is infamous for its noxious smell, which is emitted as a sign of ripeness, possibly to attract animals like wild elephants to eat the fruit and so distribute the seed. Elephants are evidently very fond of the fruit and trek for long distances to eat it. This tropical evergreen species can grow to 165 feet (50 m) and is found wild in parts of southern Southeast Asia and Indonesia. Today it is being grown extensively outside this range, notably in Thailand and most recently in Australia, to satisfy an increasing demand for a high-status, luxury fruit.

There are several different *Durio* species that provide edible fruit, but *D. zibethinus* is the only one that has been taken into cultivation. This is a connoisseur's fruit and there are many different cultivars. Aficionados tend to concur in their assessment of durian quality and there is a very high differential in prices between varieties; the best can be very expensive. Like most other commercial fruits, good cultivars are propagated by grafting. Active breeding programs in Thailand and Malaysia aim at developing varieties that lack the smell, and that can be transported without attracting complaints. Australia and China are among the countries expanding their production.

- Durian seed has very short viability, so must be sown immediately after being removed from the ripe fruit. The seed should be laid on the surface of moist compost, in the dark.
- The plant is "ultra-tropical" in its preferences and requires constant heat, humidity, and rainfall.
- A fall of temperature below 45°F (7°C) is likely to kill the tree.
- Cultivated trees require correct pruning to ensure the best possible harvests.

Eating durian fresh is the only way to appreciate its extraordinary custardlike texture and subtle flavor, which is almost impossible to describe. The smell, likened to sewage or rotting garlic, is not experienced while eating, but can be a major discouragement. The fruit is banned from hotels and on public transport in many Southeast Asian countries.

Durian timber is not of particularly high quality, although in Southeast Asia it is used for light joinery, construction, and packing-case manufacture. Even so, its reddish color is attractive enough for it to have potential as a material for furniture and interior paneling.

The fruit is heavy and covered in thick spines. Fruit falling from trees is definitely dangerous, to the extent that the trees are sometimes banned from public places.

Durian fruit is preserved in a variety of ways. Dried chips are particularly popular, eaten as a snack. Juices, cakes, candies, and a variety of desserts are also made. Durian ice-cream is widely available, and the flavor is perhaps conveyed more successfully by it than by other processed durian products.

Up to 115 ft. (35 m)

Up to 165 ft. (50 m)

*A rich, butterlike custard highly
flavored with almonds, but intermingled
with wafts of flavor that call to mind
cream cheese, onion sauce, brown sherry,
and other incongruities. The more you eat
of it, the less you feel inclined to stop.*

ALFRED RUSSELL WALLACE, QUOTED BY F. CHUNG ON DURIAN
INFORMATION BLOG

Red palm oil became [important in] supplying the caravans and ships of the Atlantic slave trade, and it still remains a popular foodstuff among people of African descent in the Bahia region of Brazil.

"PALM OIL," THE CAMBRIDGE WORLD HISTORY OF FOOD WEBSITE

Elaeis guineensis

African Oil Palm

Looking much like any other palm, the African oil palm grows to 65 feet (20 m) high. It has become one of the most important industrial crops, and a controversial one, too, yet its roots are ancient—traces of palm oil that are 5,000 years old have been found in Egyptian tombs.

The leaves are pinnate, 1–15 feet (3–4.5 m) long, and the clusters of small flowers mature to bunches of fruit, each one containing a single seed surrounded by oil-containing flesh; bunches can weigh as much as 110 pounds (50 kg). Palm oil's recent popularity as a food ingredient and a source of biofuel has led to an explosion in the development of plantations—both vast corporate ones and smallholder plots— particularly in Southeast Asia and the islands of the Indonesian archipelago. Creating these plantations has usually meant felling rain forest or draining wetland, with catastrophic effects on local biodiversity, on regional climate (because vast fires send smoke into the sky), and possibly on global climate, too.

Improving palm oil productivity is crucial to the sustainable exploitation of the plantations. One developmental goal is a dwarf palm, like the semi-dwarf grain crops that revolutionized food production during the mid-twentieth-century Green Revolution.

- Seeds are notoriously slow to germinate and normally take years.
- Commercial seed germination specialists treat them with heat, keeping them at around 104°F (40°C) for several months before sowing.
- Germination is also encouraged by treatment with hydrogen peroxide.
- The trees need humid tropical conditions throughout the year and have no tolerance of either cold or drought.

Palm oil can be processed into biodiesel, and there has been much investment in facilities for doing this. It is highly questionable as to whether this is a sustainable practice, especially given the destruction of rain forest necessary to grow the trees.

Palm oil has been a mainstay of African cooking for millennia, and in its raw "red" state it is a good source of beta-carotene. During the colonial era, the value of the oil in the manufacture of products such as margarine and a variety of processed foods was realized, and plantations established in many countries. There is evidence, however, that certain palm oil components may increase the risk of cardiovascular diseases in susceptible people.

Palm oil is used in the manufacture of soap, cleaning products, and many cosmetics. Some of these uses were first developed by traditional cultures, but have massively expanded in the current era of globalization.

Raw palm oil is used medicinally in West Africa, particularly for treating poisoning—when children accidentally drink kerosene, for example. It is also used for treating coughs and wounds, and preventing vitamin A deficiency.

Up to 40 ft. (12 m)

Up to 65 ft. (20 m)

Eriobotrya japonica ▲

Loquat

Originating in southeastern China, the loquat tree is an evergreen species, growing to a height of no more than 35 feet (10m). It is immediately recognized by its large leaves with their distinctive pattern of impressed veins and toothed edges. The main reason for growing it, the fruit, is not a practicable proposition in cold-climate zones, but in colder parts of Europe the foliage on its own has earned the tree popularity as an ornamental garden plant. Unusually, the fruit ripens in late winter or early spring, having formed from clusters of white flowers the fall before; this may be because the tree originated in an area where winters are mild. The pale orange fruit has a sweet or slightly acidic flavor, and is produced in clusters. These are somewhat difficult to harvest and often whole clusters are cut away, for later separation of the fruit prior to packing and marketing.

The loquat was introduced to Japan at least a thousand years ago, while trade routes across continental Asia took the tree into northern India and parts of the Middle East, where in some places it has become naturalized. Commercial plantations are now to be found across regions in suitable climate zones, such as the southern parts of the United States and the Mediterranean region.

- The seed from loquat fruit germinate rapidly and can be cultivated as ornamentals.
- For fruit production, grafting is used to maintain variety characteristics.
- Full sun, fertile soils, and good levels of year-round moisture are required.
- The Mediterranean and subtropical climate regions are suitable for loquat growing; trees are hardy to 12°F (-11°C), although fruit loss is likely at 25°F (-4°C).
- A great many cultivars are available.

While the fruit can be fermented into wine, in Italy the seeds are cracked open, and used to make a liqueur called nespolino. Like amaretto, nocino, and other Italian liqueurs based on nuts and kernels, nespolino is bitter-tasting due to safe amounts of cyanogenic glycosides.

Consumption of large quantities of the fruit can have a mild sedative effect. Made into a soda, they can be used as a hangover cure. They are popular in China for treating coughs (by dissolving and expectorating phlegm) and sore throats, while the leaves are used in Japan to produce a treatment for skin problems such as eczema and psoriasis.

Up to 25 ft. (8 m)

Up to 35 ft. (10 m)

Loquat fruit is widely eaten, either fresh or as a pie filling. The fruit can also be made into a sauce, stewed with spices, or used as the basis of a chutney. It can be made into a jelly because, if slightly unripe, it contains enough pectin to achieve good setting.

Loquat is not widely grown for timber but old fruit trees have value because the wood is hard, being quite similar to pear (an distant relative). It is therefore good for carving and making precision items such as rulers.

In northern and central California, where the loquat had arrived courtesy of the Manila galleons, the tree was considered an exotic ornamental.

MARIE BARNIDGE-MCINTYRE, "ORCHARD TREES OF RANCHO LOS CERRITOS: LOQUATS"

Tasmanian Blue Gum

Tasmanian blue gum is one of the most familiar species of the extremely widespread *Eucalyptus* genus. Eucalyptus trees are now found well beyond their Australian homeland, while this particular species, which is native to Tasmania and Victoria, has been especially widely planted, generally in cooler locations like those of its habitat.

An evergreen, it is capable of growing to 180 feet (55 m). It has leathery, highly aromatic leaves, and the peeling bark gives the trunk a streaky appearance that is typical of the genus. The cream-colored flowers produce large quantities of nectar, and mature to seed pods containing very small seed that is scattered over remarkable distances on the wind. The tree can naturalize, which has led to concerns over invasiveness. The growth of all eucalypts outside their homeland has led to concern that the trees lower the water table and suppress biodiversity. However, research does not necessarily confirm that they extract more water than other trees from the soil. But it is true that few other species will grow beneath and around them, because they have an allelopathic effect, as well as being very efficient competitors for nutrients and moisture. They can also be a fire risk, particularly in built-up areas, because they are full of oil that can combust explosively.

- Seed germinates best if stratified for three weeks in a refrigerator, but this is not essential to success.
- Young trees grow fast but do not tolerate transplanting, so should be grown in containers until placed in their permanent positions.
- Tasmanian blue gum in particular will survive cold temperatures, down to 10°F (-12°C).
- The tree can withstand drying, salt-laden coastal winds, but it is sensitive to drought.

Eucalyptus in general are among the world's most important sources of pulp for paper making. The huge plantations in many countries cause concern about the environmental and landscape impact of these monocultures. The building of pulp mills and the development of commercial plantations has been a particular issue in Uruguay, giving rise to border tensions with neighboring Argentina.

All eucalyptus species make good firewood, and, given their rapid growth rate, this is a particular boon to people in poor countries. The presence of good firewood trees can reduce pressure on native woodlands, too.

Producing vast quantities of nectar, eucalyptus is attractive to bees, and therefore popular with beekeepers. Eucalyptus honey has a relatively strong flavor; some find its hint of menthol a little medicinal.

Highly aromatic, eucalyptus oil is an ingredient in a number of cold remedies and balms for sore muscles. The oil from Tasmanian blue gum in particular is especially valuable because it is regarded as safe for ingestion; this is due to its lack of a harmful compound, phellandrine, that is found in other eucalypts.

Up to 100 ft. (30 m)

Up to 180 ft. (55 m)

Euterpe oleracea

Açai

The açai is a tall, slender palm that grows to 80 feet (25 m). Its pinnate leaves, 10 feet (3 m) long, help to make it one of the world's more impressive fruit trees. A species of riverbanks, wetlands, and moist soils from Central America down to the Amazon basin, açai has traditionally been the source of fruit harvested in the wild. Growing international demand for açai fruit has led to its being cultivated as a plantation crop.

The flowers and fruit are borne on shoots that emerge from the side of the trunk just below the base of the leaves. Like many palms, the tree produces extremely stiff flower and fruit stems that have a dramatically spiky appearance. Bunches of fruit typically weigh around 35 pounds (16 kg), with each darkly colored fruit being around ½ inches (15 mm) in diameter. The flavor is quite sour, even bitter, and is often compared to red wine or a hint of chocolate. The fruits are perishable and locals usually eat them within twenty-four hours to gain their maximum benefit.

- Fresh seed sown at 68–77°F (20–25°C) should germinate in weeks. For pot cultivation, several plants grown together look better than one on its own. Young plants need protection from strong sunlight.
- Constant warm temperatures and plentiful water are needed.
- Dwarf varieties are available; these start to produce fruit at only three years old.

Since around 2000, many health claims have been made for açai as a "superfood." It is full of antioxidant compounds and other nutrients, but doubt has been cast on some of the claims—particularly that it has remarkable age-retardation properties and that it can cause rapid weight loss.

Up to 50 ft. (15 m)

Up to 80 ft. (25 m)

The entire fruit, including the pulp and skin, is eaten by some Amazonian tribespeople, for whom it is an important part of the diet. Açai fruit can be made into a thick pulpy drink with a strong purple color, which has led to it being freeze-dried and marketed internationally in smoothies.

Açai's large leaves can be used for making waterproof roofing for houses. They are tied to poles that are then laid horizontally, the leaf surfaces pointing downward and overlapping in the way of tiles.

Fagus sylvatica

European Beech

This deciduous forest tree is found in either mixed forest or natural monocultures across Europe north of the Mediterranean region, and at higher altitudes in southern Italy. Growing to 150 feet (45 m), with a majestic branching habit and a smooth gray trunk, it is arguably Europe's finest tree for landscape planting. The leaves are extremely efficient at capturing light, making the floor of beech woods dark for other plants and suppressing their growth. Historically, beech forest would have covered much of the more fertile soils in central Europe, but over the millennia many of these were felled. More recently, cessation of animal grazing has resulted in the recovery of beech forest, but often at the expense of wider biodiversity.

- Beech nuts need to be sown in fall as soon as they are ripe, and will germinate the following spring if the winter has been cold.
- The trees should be transplanted within two years, and planted out in full sunlight on a well-drained fertile soil.
- Beech takes clipping well and is often used for hedging.

In central Europe, twigs were formerly cut and stored so that their leaves could serve as forage for cattle in winter when other forms of nutrition were scarce.

The wood is very popular for furniture making, because it is light in color, with a dense, short grain that makes it easy to work; its very attractive finish makes it popular in kitchens. That it can be steamed and bent into shape has led to its inventive use in "bentwood" furniture.

Beech nuts, known as "mast," have traditionally been used to feed pigs, with the animals allowed to run free under the trees. Humans can eat mast, too, but the skin has a high tannin content so should be removed and the nuts toasted like chestnuts.

Up to 165 ft. (50 m)

Up to 150 ft. (45 m)

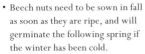

Ficus carica

Fig

The common fig is a familiar fruit tree of moderately warm climates, native to the Mediterranean area and the Middle East. Growing to no more than 35 feet (10 m), it has very distinctive, deeply lobed leaves, leading to its being grown as an ornamental in its own right.

It has a unique floral structure, like all figs, with the flowers being invisible and buried inside an enclosing structure. This has a small hole at one end that is penetrated by a very small wasp that pollinates the flowers and lays its eggs in the fruiting structure. In this species, only hermaphrodite flowers play host to the wasp; female flowers turn into fruit that is completely free of wasp larvae.

- Figs are propagated by cuttings, grafting, or layering, where a stem is pegged into the ground to encourage rooting.
- The tree is hardy in temperatures down to around 5°F (-15°C), although warm summers are necessary to ensure that the fruit will ripen.
- The trees are naturally vigorous and need a confined root run, as well as pruning, to stimulate fruiting.
- Fruit takes nearly a full year to ripen.

Given that the fresh fruit is delicate and does not transport well, there is a tendency for it to be used as an ingredient for making products with a longer shelf life. One such is the "fig roll," a form of cookie where a mass of cooked fig is enclosed in a pastrylike wrapping.

Historically, unprepared figs have been recommended as a laxative due to their high fiber content. Foods with high fiber content can reduce hunger cravings.

Fresh figs are a juicy delight, but historically, given their short season, much of the harvest was dried. Dried fruit keeps well; formerly it was traded far beyond its source area.

Up to 35 ft. (10 m)

Up to 35 ft. (10 m)

Ficus religiosa

Pipal, Sacred Fig

One of many tropical species of fig, this species is widespread across southern and southeastern Asia, both naturally and through deliberate planting. Growing to 100 feet (30 m), it is semi-evergreen, which means that it may drop its leaves in the dry season. The leaves have a distinctly long point at the tip, which has evolved to help water to drain from the leaf surface.

The tree is of considerable religious significance in Asia. Although treated as sacred by Hindus, it is of especial importance for Buddhists because the Buddha himself famously achieved enlightenment while seated under the tree about 2,500 years ago. That tree stands beside the Mahabodhi temple in Bodhgaya, India. Siddhartha Gautama sat under a pipal tree, resolving not to move until he attained enlightenment. Ancient specimens in temple courtyards, said to be descended from that original tree, are known as "bodhi" or "bo" trees.

Great use is made of the fruit in the Indian medical system of Ayurveda, with a huge range of complaints being treated by a variety of preparations. Clinical trials have shown that the fruit does have anti-microbial and analgesic properties.

Up to 35 m (115 ft.)

Up to 30 m (100 ft.)

The leaf's tip gives it an elegant appearance, which is made use of in the traditional art of leaf painting. Leaves are prepared by allowing them to soak underwater to decay their flesh. Images are then painted onto the resulting fine network.

In India, a home remedy for sickness and diarrhea consists of a pickle made of pipal leaves preserved in mustard oil. For a medication it is very popular and surprisingly tasty.

- Seeds germinate at 68–77°F (20–25°C) over a period of several months.
- Sacred fig is also easy to propagate from cuttings.
- The species needs warm temperatures, with a minimum of 54°F (12°C).
- Loam is the best soil. When planting, use soil with a pH of seven or below.
- This fig naturally grows in a forest environment, so direct sun can be damaging to leaves.
- The plant is popular as bonsai, as it forms gnarled shapes very easily, with aerial roots developing from the branches.

Fraxinus excelsior

Ash

One of the most common and widespread deciduous trees of Europe, found as far east as the Caucasus, ash forms a high proportion of the biomass of woodland across the continent. Growing to 115 feet (35 m), it is both a pioneer species and an important and potentially long-lived component of mature forest. It is almost instantly recognizable by its pinnate leaves, the smooth gray bark of younger trees, and, in winter, its dark buds. Its large output of seeds is a major food source for birds and small mammals.

Although found on many soil types, ash is particularly associated with calcareous ones. It is tolerant of a short and cool growing season, which helps account for its being widespread in areas such as northern England and Scandinavia. It is common partly because it has a remarkable tolerance of air pollution, which enabled it to survive and spread in the 150-odd years after the start of the Industrial Revolution, when pollution was very severe. Traditionally, it was one of the trees often grown as pollard or coppice, to supply poles that could be turned on a lathe, or drawn with a knife; but, unlike other coppice systems, ash coppicing has seen relatively little revival.

- Ash seed germinates readily but only after a long period of dormancy that comprises two cold periods
- Unripe seed will germinate more quickly than ripened seed.
- Seedlings grow rapidly, and should be planted out in full sun in a moist but well-drained soil.
- Ash has a very strong tendency to grow upright. This makes it unsuitable for pruning or hedging, which requires lateral growth.

The most common traditional uses of ash have been for furniture and tool handles. Because it splits cleanly, ash has been widely used for a form of basket making where a more robust material than willow is required. Hurley sticks, used in the heritage Irish sport of hurling, are made from ash—thus, the game has been called "the clash of the ash." Eastern Europe supplies most of the ashwood used for such purposes.

The wood is dense, hard, and shock-resistant, making it useful for a wide variety of purposes. Early aircraft used ash for their framing, and it was used for parts of the British de Havilland Mosquito in World War II to help reduce its weight. Ash also features visibly in the body frame of the Morris Traveller, one of the most successful British cars of the 1950s, and one still very popular with enthusiasts.

Up to 100 ft. (30 m)

Up to 115 ft. (35 m)

A traditional Swedish way of making a tree house is to pollard an ash, put a wooden platform on the stump and then let the new growth grow up and around it. The Swedes also used to plant sacred ash trees in the center of their properties.

Ash makes very good firewood because it will burn while still green and needs the minimum of seasoning. Sometimes it is mixed with beech to damp that wood's tendency to spit embers.

> *[Ash trees] secrete a sugary substance from their bark and leaves, which until the early part of this century was harvested and sold under the name "manna."*
>
> DARL J. DUMONT, "THE ASH TREE IN INDO-EUROPEAN CULTURE", *MANKIND QUARTERLY*, SUMMER 1992

Both Sargent and Wilson of the Arnold
Arboretum expressed the often-quoted
opinion that ginkgo was probably
extinct in the wild and that it was
saved from total extinction by
Buddhist monks who cultivated it in
the gardens surrounding their temples.

PETER DEL TREDICI, "WHERE THE WILD GINKGOS GROW,"
IN *ARNOLDIA*, 52:4 (1992)

Ginkgo

The ginkgo is an ancient and deciduous distant relative of the conifers, a survivor from the days of the dinosaurs and the sole survivor of a once common family. It has been extensively cultivated in China, Korea, and Japan for two millennia, partly because of a close relationship with Buddhism; as a consequence it is very difficult to determine where it is growing naturally. Recent research has shown that some trees near the city of Hangzhou in eastern China are almost certainly natural.

Ginkgo leaves are very distinctive; in fact, they are unique. They cluster on what appear, on young plants, to be rather gawky branches. The tree grows fast, with time forming a distinguished-looking tree that grows to 100 feet (30 m). It also builds up a massive stump, from which it can regenerate after cutting or fire; some of the Hangzhou trees have these, making it not only highly likely that they are natural but also very difficult to determine their ages. This characteristic has no doubt helped the species survive over its very long period of geological time.

Male and female ginkgo trees grow separately. The fruit, borne on females, smells very unpleasant, so males tend to be selected as street trees, for which they are more suitable anyway, given their more limited spread. A variety of cultivars now exist, and these have largely been chosen on the basis of the shape of mature trees.

- Seed should be sown in fall or stratified for at least a month before planting at a temperature of 68°F (20°C).
- Seed germination will be slow and erratic.
- Young trees grow quickly and are tolerant of a wide range of soil types and climates. However, they prefer fertile and moist but well-drained sites.
- The trees do not respond well to pruning.
- Ginkgo is popular as a street tree because it tolerates atmospheric pollution.

The nuts can be eaten and are included in a variety of recipes from Japan and China—the Chinese call them "silver apricots"—particularly those that are associated with Buddhist vegetarian cuisine or are eaten on special occasions, such as weddings, where certain foods are seen as auspicious. They also go into sweet soups, such as one made with boiled tofu skin.

Research has indicated that ginkgo nuts or ginkgo-derived dietary supplements can have an impact on short-term memory and attention span. This benefit is thought to derive from its effect on blood circulation, and so healthy people are unlikely to see memory improvement.

Widely used in traditional Eastern medical systems, ginkgo has recently received much publicity as a possible treatment for dementia and Alzheimer's disease. Sadly, researchers have not found firm evidence to support this claim.

Recent research, based on the reactions of laboratory animals to dietary ginkgo, indicates that consumption of ginkgo products on a regular basis may cause health problems. The leaves may even be carcinogenic in their effect.

Up to 46 ft. (14 m)

Up to 100 ft. (30 m)

Gliricidia

With pinnate leaves and pink pea flowers, this is a medium-sized tree that grows to a maximum height of 40 feet (12 m). Its appearance announces that it is a member of the pea family. It has become a very important species in agroforestry, and is now grown in many warm-climate regions far from its original home in Central America. It is adapted to a seasonally dry climate and drops its leaves during drought.

The tree's main use to humankind is as a forage crop, for which it has proved very popular because of its fast growth rate and adaptability to a wide range of soil types. It can flourish in very infertile soils because, like nearly all "peas," its roots play host to nitrogen-fixing bacteria. There is evidence that it can suppress the growth of other plants through allelopathy, although possibly only in certain conditions.

- Propagation through seed or cuttings is easy, with high percentages of both germination and plants grown from cuttings.
- Young plants grow rapidly and can even flower in their first year in warm climates.
- While it can be an invasive species in some locations, cattle and other grazing animals find it very palatable and so easily keep it in check.

As a nutritious and palatable forage crop for ruminants, gliricidia is very popular with farmers across the developing world. Trees may be grown scattered through grass or other crops, and the foliage cut for feeding animals. Non-ruminants do not find it so easy to digest, however.

The bark and the fermented leaves produce a toxin, which can be used as a rodenticide.

Since a stem thrust into the ground in the rainy season will soon root, this has proved a popular species for making living fences.

Up to 35 ft. (10 m)

Up to 40 ft. (12 m)

Unfortunately no autopsy has been performed on a rat killed by gliricidia.

LAURA S. MEITZNER, MARTIN L. PRICE
AMARANTH TO ZAI HOLES (1996)

Lignum Vitae

Native to the Caribbean region, lignum vitae is a slow-growing tree that eventually reaches 35 feet (10 m) in height. It is now on the endangered list because of a long history of over-exploitation. The wood is immensely hard, and so dense that in water it sinks like a stone. With round, evergreen leaves and purple flowers that fade almost to white, it is an attractive plant and thus occasionally grown as an ornamental. Jamaica helped to raise its conservation profile internationally by choosing it as its national flower.

Historically the species formed part of a dominant vegetation community on some Caribbean islands. Following its colonial exploitation it is most likely to be found on poorer soils; here, its slow growth allows it to compete with species that, on more fertile ones, would grow at a faster rate and shade it out.

A resin from the wood has been used by traditional medical systems for treating various diseases, including syphilis. It also has anti-inflammatory properties and is included in medications for sore throat, gout, and rheumatoid arthritis.

Such a hard wood used to be popular for croquet mallets, bowling balls, mortar and pestles, the handles of smaller specialist tools, and other applications requiring a combination of hardness and weight—including police batons.

The gum is still used today in the preparation of test cards for a potentially life-saving medical test, used to detect blood in stools, a sign of possible colonic cancer.

• Fresh seed germinates relatively quickly, but erratically.
• Seedlings grow slowly initially, but take off at a faster rate after they reach 8 in. (20 cm).
• The tree survives short-lived frosts but its ideal climate is a warm one, with a notable dry season.
• Plants do not transplant well, so starting seedlings in pots is preferable to planting them in the ground initially.
• The presence of very light overhead shade helps to reduce drought stress in young plants.

Up to 40 ft. (12 m)

Up to 35 ft. (10 m)

Kentucky Coffee Tree

This species from the central Midwest of the United States grows to 65 feet (20 m) and makes a fine ornamental specimen. However, it is not often planted because it is one of the first trees to lose its leaves in the fall, and one of the last to produce leaves in spring and having relatively few twigs, it has the appearance of a dead tree for much of the year. The pinnate leaves can be as long as 35 inches (90 cm), emerging pink before turning green for the summer; in fall their color is yellow.

The seed pods need to be disrupted physically before they can release the seeds, which means that the tree tends to be restricted to wet areas, where the pods rot before the seeds lose viability. It has been suggested that they evolved to be eaten by a now-extinct animal, such as a mastodon. Although the tree is found across a wide range, it is regarded as rare.

- The seed needs to be damaged with a file to help moisture gain access, or soaked in sulfuric acid. Germination is then rapid.
- The tree requires full sun, and does best on moist, calcareous soils.
- Its very large leaves made up of smaller leaflets make it a good shade tree.

The hard seeds were used by some Native American tribes as beads in jewelry, as a component of a rattle (with the seed enclosed by a gourd), and as counters or dice in games, both recreational and ritual.

Early settlers ground up the seeds as a coffee substitute. The seeds contain toxins but these were neutralized by the roasting process. A laxative tea can be made from the leaves.

Human poisonings by the coffee tree are virtually unknown, but cattle have died after eating the leaves or drinking water in which fallen leaves have dissolved.

Up to 30 ft. (10 m)

Up to 65 ft. (20 m)

Hamamelis virginiana

Witch Hazel

Growing to around 20 feet (6 m), this tree is an understory species of deciduous forest and woodland-edge habitats in eastern North America. The fact that it is shade-tolerant and so can thrive under a mature forest canopy means that it can be used as an indicator of whether a forest is old growth or mature second growth. The strange little flowers, which look like scraps of yellow plastic shavings, appear in fall and fill the air with a sweet fragrance for some distance from the plant. The small fruit take a year to mature; on doing so, they shoot out their tiny seeds for distances of up to 23 feet (7 m). The leaves are coarse and broadly oval, with a great similarity to those of the Old World hazel, hence the second part of the name.

A fluid, itself known as "witch hazel," is derived from the leaves and bark and is still approved for medical purposes. It has been shown to reduce inflammation, draw tissue together, and slow down bleeding—all useful properties in dealing with minor cuts, insect stings, and inflammation.

Up to 20 ft. (6 m)

Up to 20 ft. (6 m)

The younger branches have been used by water diviners, probably because the foliage of the tree is similar to that of the hazel (*Corylus avellana*), which is favored by water diviners in Europe.

Witch hazel, distilled or in the form of a gel, is highly regarded as a treatment for the effects of poison ivy—a species that often grows alongside it in woods.

- Witch hazel seed is notoriously slow to germinate, and even fresh seed requires eighteen months in the ground before it comes up, during which time it will have been chilled by two winters.
- Commercially, plants are often sold grafted onto more robust rootstock.
- Growth is slow, and there is a need for moist but well-drained soils, preferably acidic and humus-rich.
- The tree is naturally tolerant of partial shade.

Ilex aquifolium ⚠

Holly

Holly is one the most widespread of central and northern Europe's rather limited evergreen flora. Potentially growing to 50 feet (15m), it is more usually seen as an understory shrub in woodland, in hedges, or in gardens. Its ecological history is somewhat mysterious, because in the British Isles at least it once formed woods, but virtually none of these remain.

Male and female flowers are found on separate trees, with only the females bearing the bright red berries that are regarded as an essential part of Christmas decorations. Numerous horticultural varieties have been selected for their foliage, productivity in berries, or having yellow berries.

A popular Christmas decoration, holly has also traditionally stood in for palm leaves on Palm Sunday in Germany; the tree is even called *Stechpalm* in German.

Surprisingly, holly was once widely used as forage for livestock, with the less spiny leaves near the top of the tree being cut to provide winter feed. This may partly explain the presence of holly in traditional hedgerows.

Tight-grained, well-seasoned holly wood is good for carving and sanding to a smooth finish. It has a pale appearance, almost like ivory, although it darkens on exposure to air if not varnished or oiled.

- Holly seed is deeply dormant; germination may occur in the second spring after sowing. Otherwise, forty weeks at 68°F (20°C) followed by twenty-five weeks at 39°F (4°C) will begin germination.
- Named cultivars are propagated by cuttings.
- Plants require a reasonably fertile soil in sun or light shade.
- Holly takes pruning well and can be trimmed into topiary.

Up to 35 ft. (10 m)

Up to 50 ft. (15 m)

In many parts of Britain it is believed that cutting down the holly brings misfortune . . .

ARCHIE MILES, *THE TREE IN BRITAIN* (1999)

Ilex paraguariensis

Yerba Mate

Sometimes growing to 50 feet (15m), this evergreen tree from Paraguay and the area of Argentina's border with Brazil is usually seen as a shrub. It is the source of a herb that is of major importance locally but that is little more than a curiosity elsewhere. The leaves have the serrated edge typical of other *Ilex* species, but they are not spiny. Like other hollies, it has red berries.

Plantations of the tree are mostly in Brazil, with the remainder being in Paraguay and Argentina. The leaves contain caffeine, also referred to as a mateine, and various alkaloids. They are harvested and dried for the preparation of mate, or *chimarrão* as it is known in Brazil.

- Seed has to be lightly scarified before sowing, at around 68°F (20°C); germination may occur within months or it may be necessary to wait much longer.
- Growing conditions need to be frost-free, preferably subtropical, and with the benefit of light shade.
- The plants adapt well to pot cultivation, although leaf production is reduced in potted specimens.

As a drink, mate is a mild stimulant, much like coffee, that is associated with a great deal of social ritual and requires complex equipment to make. A large number of leaves are steeped in hot water, which is drunk through a special tube with a filter at one end. The supply is then replenished, often from a vacuum flask.

A correlation has been noted between mate drinking and oral and other cancers; however, the high temperature at which mate is often drunk could be the cause of the cell damage, rather than the mate itself.

Up to 35 ft. (10 m)

Up to 50 ft. (15 m)

Research indicates that mate helps to break up fat molecules in the bloodstream; long-term use may benefit some people.

Juglans nigra

Black Walnut

Reaching a maximum height of 130 feet (40 m), this deciduous tree tends to grow mostly in river valleys throughout the Midwest and eastern regions of North America, often in the company of American elm (*Ulmus americana*). The pinnate leaves are 1 to 2 feet (30–60 cm) long with up to twenty-three leaflets. The species tends to occur as single specimens in woodland because it cannot tolerate the shade of other trees. Nuts can be borne by surprisingly young trees, but full production only begins after twenty years; the nut production is irregular, with many trees only producing a good harvest every two or three years.

Along with many other species, the black walnut was brought from America to Europe in the seventeenth century, whereupon it was occasionally grown as a timber and nut-producing tree. Like other walnuts, it is strongly allelopathic, suppressing the growth of other plants within its root zone, but appreciation of its high-quality timber has led to its being frequently planted. The allelopathic compound within the tree, juglone, becomes activated only when outside the plant, when it is activated by contact with oxygen. The adverse effect is not restricted to plants—juglone is also a deterrent to insects.

- Nuts intended for planting should be de-husked and then tested for viability by immersing in water—plant only those that sink.
- Autumn sowing or stratification will be necessary, with protection against animals likely to dig up the nuts.
- Deep, fertile, moist soils in full sun are required.
- Dry sites and frost hollows should be avoided for planting, but there is some tolerance of flooding.

Black walnut is the most highly regarded North American timber, being dense, very resistant to shock, easy to work, and very good looking. It is in great demand as a material for furniture, worktops, gunstocks, and flooring. Poaching of the trees is a problem, but identifying illegally harvested timber can now be aided by DNA profiling, where logs offered for sale can be matched to the stumps of stolen trees.

Walnuts are an excellent and tasty source of protein, playing a part in many cuisines, and the nuts of *Juglans nigra* are reckoned by some to be superior in flavor to those of the Old World walnut, *Juglans regia*. Walnut oil is an alternative to olive and other oils for use on salads, having a warm, nutty flavor. It is used less often in cooking because it is relatively expensive and tends to lose its flavor when heated.

Up to 100 ft. (30 m)

Up to 130 ft. (40 m)

Walnut husks produce a dark brown dye, notorious for staining the hands of harvesters. Tannins act as a mordant, and it has been used for dyeing hair, for fabrics, and as a wood stain.

In the past, artists would use walnut oil for oil painting because it dries quickly and does not yellow with age. Nowadays walnut has largely been replaced by linseed oil, but debate continues over which is superior for the purpose.

If you want to plant walnuts, take two bushels of nuts into the forest—one for the squirrels to bury and eat later, and one bushel for them to bury, forget about, and let grow.

FORESTER QUOTED BY THE
UNIVERSITY OF MINNESOTA
EXTENSION SERVICE WEBSITE

Juniper

This is the most widely distributed of all woody plants, being found in northern temperate and mountain regions of Eurasia and North America. Typically a plant of exposed poor soils, especially calcareous ones, it often forms low-growing, gnarled shrubs as a response to wind exposure. It is capable, however, of forming a tree 35 feet (10 m) high. The foliage—very spiny needles—offers a good defence against predation, and also protects nesting birds. Rarely for a conifer, it produces berries; these take a long time to ripen and are visible on the plant for much of the year.

As might be expected from its being so widely distributed, there is considerable variation, and several subspecies are recognized by botanists. A number of cultivars, selected chiefly for a distinctive habit, are widely used in horticulture.

- Propagation by seed is difficult and unpredictable, with seeds taking as much as five years to emerge.
- The use of cuttings is common commercially.
- Plants are very hardy and stress-tolerant, and thrive on thin and poor soils.
- Full sunshine is essential.
- Cultivars are often planted in gardens, with shapes varying from slender and upright to sprawling.

Where the harvesting of timber is a practical proposition (not all the plants grow into trees), it has been used for small items, such as containers. The gnarled shapes that juniper can form, and its pleasant resinous aroma, make it ideal for souvenir carvings. Historically, juniper was used for making treenails, the tough pegs used in the construction of wooden ships.

The bitter, strongly flavored berries have long been used to season meats, especially game birds and venison.

Distilled spirit has long been flavored with juniper berries and sold under names such as gin, genever, and borovička.

Up to 25 ft. (8 m)

Up to 35 ft. (10 m)

Kigelia africana

Sausage Tree

One of the plant world's more bizarre sights is that of this tree in the fruiting season; each fruit, technically a berry, is sausage-shaped, up to 3 feet (1 m) long and up to 6 inches (15 cm) wide, almost woody in texture, and hangs from a cordlike stem. The tree itself is substantial, growing to a maximum height of 65 feet (20 m). It is an evergreen species in areas with moist climates but drops its leaves during dry seasons elsewhere. The dark red-purple, trumpet-shaped flowers are pollinated by a wide variety of insects.

The sausagelike fruit contains a mass of pulp that is eaten by monkeys and birds such as parrots. Those fruits that fall off the tree or are knocked down by animals are eaten by an extensive range of mammals, which distribute the seeds in their dung.

The unusual appearance of the flowers and fruits has led to the sausage tree being planted as a curiosity in botanical parks and gardens a long way from its large natural range in tropical Africa.

Extracts made from sausage fruit are used in traditional medical preparations for the skin, and there is good scientific evidence that they reduce the impact of a variety of conditions, such as psoriasis, acne, and eczema. Specialist cosmetics companies are beginning to use it in everyday products that promote skin health, too.

The wood is exceptionally strong yet easy to work and produces a good-quality timber for use in construction and furniture making. The large size of the tree also makes it suitable for making dugout canoes and oars. The wood is also used to make ox yolks.

- Seed germinates within a few weeks at 68–77°F (20–25°C).
- Young plants are fast growing and must be placed carefully—the tree canopies and root systems are extensive.
- Falling fruit can cause damage or injury.
- Established trees will tolerate temperatures as low as 39°F (4°C).

In Malawi the seeds are roasted and used in beer making as an aid to fermentation. A traditional brew called "muratina" includes honey.

Up to 80 ft. (25 m)

Up to 65 ft. (20 m)

Larix decidua

European Larch

Larches are unusual in that they are conifers and yet deciduous. The European larch is a distinctive part of the landscape of the European Alps and the Carpathians for its grass-green foliage, which turns yellow in fall. The tree is also commonly planted elsewhere as a source of fast-growing timber; the larch can potentially grow to 130 feet (40 m).

The tree is extremely cold tolerant, surviving temperatures down to -58°F (-50°C), its leaf-dropping habit enabling it to endure winter conditions extreme enough to damage the evergreen foliage of other conifers. Young trees are upright and elegant with upwardly swept branches, while older specimens broaden out. The larch is a pioneer species and does not mix well with other species.

- Seed should be sown in fall or given six weeks of stratification; it germinates rapidly and well.
- Young plants should be set out in positions with full sunlight, preferably on fertile and deep soils.
- Larches are naturally stress-tolerant and can cope well with poor, stony soils, but are intolerant of poor drainage or drought stress in warm summers.

Larch is one of the best materials for domestic fencing. It is easily made into thin, flexible strips that can be assembled into strong but lightweight panels that offer almost total privacy.

Up to 50 ft. (15 m)

Up to 130 ft. (40 m)

Larch wood is flexible and easily split, making it ideal for a wide range of purposes. High-quality material is used for building yachts and other wooden vessels; such timber must be knot-free, so it is obtained from older trees that have been pruned continuously.

Fresh young shoots have a lemony flavor and make a good garnish for vegetables or soft cheeses, or even for creamy desserts.

Laurus nobilis

Bay Laurel, Sweet Bay

The evergreen bay laurel is only debatably a tree because it tends to form a large multi-stem shrub that grows as high as 60 feet (18 m). Although highly familiar as a cultivated plant, it is limited in its natural habitat to scattered pockets ranging from Syria across to Spain, Morocco, and the Canary Islands. The pockets are the remains of vast forests of evergreen trees and large shrubs that covered the region in the moister climate of the Tertiary era. Not surprisingly, the species is very variable and shows considerable genetic variation. In cultivation, it is often grown to be heavily clipped and shaped as topiary.

Bay leaves are among the fundamental ingredients of Western cuisine, and were introduced from the Mediterranean region. Easily dried, they maintain their flavor for up to two years. Usually they are added to a stew or sauce for background aroma.

Bay can be processed to produce an essential oil that is used in massage and to treat rheumatism. Bay leaves can be used to make a salve to treat irritation caused by poison ivy or stinging nettles.

Up to 30 ft. (10 m)

Up to 60 ft. (18 m)

Laurel oil is included as an ingredient of Aleppo soap, a traditional soap mainly made from olive oil in the city of Aleppo in Syria. Aleppo soap is regarded as highly beneficial for skin problems such as dermatitis.

- Autumn sowing or stratification of around nine weeks is needed for germination.
- Young plants are easy to grow on, but shelter from hard frost and strong or cold winds is needed.
- The species is hardy to around 10°F (-12°C); well-established plants will recover from the base if top growth is killed.

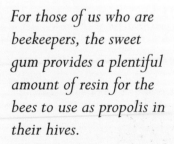

For those of us who are beekeepers, the sweet gum provides a plentiful amount of resin for the bees to use as propolis in their hives.

"BOHOREX," COMMENT ON THE SELF SUFFICIENT GARDENER WEBSITE

Liquidambar, Sweet Gum

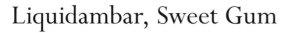

With its palmate leaves, the liquidambar tree looks just like a maple at first, but closer examination reveals that the leaves are alternate, whereas maple leaves grow opposite one another on the twig. The sweet gum is a deciduous species that will grow as high as 130 feet (40 m). The *Liquidambar* genus goes back a long way in the fossil record and was once very widely distributed; today, the sweet gum is the only New World species. Native to the eastern United States and parts of the Mexican highlands, it is a second-stage pioneer species, often being the first deciduous tree to grow alongside pines on cleared land. It is increasingly intolerant of shade as it ages, and so does not form a significant part of mature forest. It seeds aggressively, and also regenerates after felling or other damage by throwing up suckers from the base or the roots.

The tree's vivid fall color, and the tendency of some cultivars to have a narrow shape, has gained it popularity as an ornamental tree; cultivars are selected primarily on the basis of leaf color and tree shape. The flowers are insignificant but the fruit is very noticeable, each one a perfect globe with an array of spines. A distinctive feature is the corky ridges on the twigs—an adaptation that awaits an explanation.

- Seed should be sown in fall, or stratified for two months, after which germination is usually relatively quick.
- Young trees need full sunlight and moist fertile soils.
- The species is notably tolerant of occasional flooding and the tree is found in drier swamps.
- If liquidambar is being grown for its fall color, a named cultivar with predictable characteristics should be purchased.

Despite often warping and not being very strong, this is an important timber in the United States because its high rate of growth makes it a good plantation tree. The wood is an attractive reddish-brown, and so is used for veneers and very extensively in plywood—in which form it appears in a wide range of furniture and home products. Commercially, it goes under the somewhat misleading name of satin walnut.

The reddish resin, known as storax, was used by Native Americans and pioneers as a form of chewing gum. Often it was enough to scrape back the bark and collect the resin that oozed out. Extractions from the resin have a variety of other uses, including as a fixative with a grasslike aroma in perfumery.

Mixed with tobacco, and included probably because of its pleasant aroma, liquidambar gum is known to have been used by the Aztecs as a smoking material and as incense. The gum was supplied to the Aztecs by producers in southern Mexico.

In medicine liquidambar gum is known as copalm or copalm balsam, and it has been used in a range of products for treating chest complaints. Among Native Americans, it was also used to treat dysentery and sciatica.

Up to 80 ft. (25 m)

Up to 130 ft. (40 m)

Tulip Tree, Tulip Poplar

Soaring to 200 feet (60 m), often with a trunk clean of branches, a mature tulip tree is a magnificent sight. Common east of the Appalachians in the United States, it is a pioneer species and rarely found in mature forest. Fast growing, but also long-lived, it is eventually displaced by oaks and hickories. While it thrives on moist, fertile soils, it is not a marshland tree. The flowers have only a vague resemblance to tulip flowers but the tree is distinguished, at least in summer, by its unique saddle-shaped leaves, which turn to golden yellow in fall. Unusually for a North American forest tree, the flowers are insect pollinated. The tree has been one of the most successful American introductions to western Europe.

- Seed should be sown in fall or given three months stratification; germination may be erratic.
- The trees need to be planted out in full sun or light shade in moist, fertile soils, where they will make rapid growth.
- Drainage should be good, but occasional flooding is tolerated.
- Branch wood tends to be weak, so snow or wind damage is possible.

The flowers produce copious amounts of nectar and are therefore very attractive to bees. Undiluted by other nectar sources, however, tulip tree honey is too strongly flavored for many people.

American tulipwood, as it is known commercially, being easy to work and strong enough for most purposes, was very popular with both Native Americans (it was ideal for making dugout canoes) and the first settlers. Resistant to termites, it has been used for making cladding. Often having an attractive pink color, it is used in furniture making.

Up to 80 ft. (25 m)

Up to 200 ft. (60 m)

Modern usage of the timber is overwhelmingly for pulpwood. Fast-growing and one of the largest trees in the eastern United States, the tulip tree is used for reforestation.

Macadamia tetraphylla

Macadamia

An evergreen tree that can grow 60 feet (18 m) high, macadamia is the only member of the Protea family to be grown for food. Its pendant heads of creamy flowers ripen to produce the familiar spherical nuts. Macadamia was the first Australian food plant to be developed commercially, with the first plantation being started in the 1880s. Since that time the tree has been introduced to many countries beyond Australia, with production in Mexico becoming particularly successful.

Ironically, the tree has become very rare in the wild. It is native to a small area bordering Queensland and New South Wales, and the wild populations are thought to be reduced to a few hundred individuals. There are other *Macadamia* species, but the fruit of some contain cyanogenic glycosides; native peoples have learned to remove them with leaching.

Macadamia wood is difficult to work but highly decorative, which makes it suitable for small, high-value projects. Given the growing popularity of the nuts, an increasing amount of timber is likely to become available.

Macadamia nuts contain ingredients that may help reduce the ill-effects of ageing on the skin; consequently, they are used in the manufacture of some cosmetics. However, like all nuts, they can produce an allergic reaction in some people. It is advisable to perform a skin test with a tiny amount of the cosmetic before adding it to a skin-care regime.

Up to 40 ft. (12 m)

Up to 60 ft. (18 m)

With their rich, buttery flavor, the nuts are high in fat, and relatively low in protein—a luxury food. But of all the *Macadamia* species, only the nuts of *M. tetraphylla* and *M. integrifolia* are safe to eat raw.

• Shop-bought nuts will never germinate.
• If macadamia seeds can be obtained, they need to be filed before sowing. Germination may take as long as six months.
• The trees are easy to grow in any well-drained soil.
• The climate should be frost-free but not generally exceeding 77°F (25°C).
• Nut production normally starts in four years.
• Ideally, named cultivars should be sought to ensure the best results.

A seed hidden in the heart of an apple is an orchard invisible.

WELSH PROVERB, QUOTED BY JONATHAN SILVERTOWN IN *AN ORCHARD INVISIBLE —A NATURAL HISTORY OF SEEDS* (2009)

Apple

Bearer of the world's most popular fruit, the apple tree is not large, growing to 15 feet (4.5 m), and is relatively short-lived—between 100 and 150 years. Originating in the border area of Kazakhstan and Kyrgyzstan, it was introduced into cultivation millennia ago, and distributed along the trade routes of central Asia. The majority of today's apples are produced not that far away from there, in China.

The apple exhibits extraordinary genetic diversity—there are now some 7,500 varieties—but of these only a tiny proportion are today commercially viable; enthusiasts seek out and grow many of the rarer ones. Apple diversity received a boost when the tree was taken to the European colonies in America in the seventeenth century. After the importations a frenzy of genetic mixing occurred when large numbers of trees were raised from seed.

Apple trees themselves would probably be popular ornamental plants even if their fruit were not so desirable because in spring they are covered in attractive flowers that have the almost unique quality of opening pink and fading to near white. Dwarfing rootstocks (initially brought from central Asia by the ancient Greeks) make it possible to enjoy these flowers and their fruit in all but the smallest gardens.

- Apples need to be grafted to maintain their parental characteristics.
- Ideally the trees should be grown in full sun on deep, fertile soils, with good drainage.
- Avoid sites that experience late frosts.
- Apple production depends on skillful pruning to maximize the number of potentially flowering shoots.
- Dwarf varieties may be short-lived but crop earlier and younger than ones on non-dwarfing rootstocks.

Apples are the quintessential healthy fruit, and their ability to store well and be transported without too many problems is an essential part of their success. Apples are now processed in myriad different ways, with apple juice in particular being an almost universal commodity. Apples are generally used in sweet dishes that make the most of their texture and flavor, with North American and central European cuisines being particularly inventive.

Somewhat surprisingly, apples have few applications in traditional medical systems. A tea made from apple leaves or bark has been used for treating digestive problems, and is antiseptic. Pectin, the main chemical extract, is used in the food industry, and in making cultures for laboratory work.

Apple wood is hard and not always easy to work, although it turns well and has an attractive appearance. It is used by amateur woodworkers to produce decorative pieces.

Apple wood has a good reputation as a firewood, burning long and hot, and with an aromatic fragrance that is popular for smoking foods. However, relatively little is sold for burning.

Up to 10 ft. (3 m)

Up to 15 ft. (4.5 m)

Mango

One of the world's favorite fruits, the mango is borne by an evergreen tree native to India and Southeast Asia. Growing to 130 ft. (40 m), mango trees can live a long time; very old specimens can be a much-loved feature of their local landscape.

Mangoes vary considerably, not least in flavor, which is basically sweet but with a number of aromatic tones. Their season tends to be brief, especially when it is cut short by the dry season, as in India. Although India is the largest producer, hardly any are exported, such is the national passion for eating them. Historically, trees could only be propagated by seed, so there was a great deal of variation; this problem is now overcome by the grafting of select cultivars. The popularity of the fruit is growing, with an increasing amount exported by countries with highly developed agricultural industries, such as Brazil.

The most esteemed variety in India is called the "Alphonso"—the name commemorating a sixteenth-century Portuguese naval commander who played a part in establishing Portuguese colonies on India's western coast. Modern mango breeders have to focus on shelf life and ease of transport as well the traditional criterion of flavor.

- Planting a mango stone in compost in a warm place is a simple way to get started.
- Quite often several seedlings emerge—one should be selected and the rest cut off.
- Growing on requires constantly warm conditions, with a minimum temperature of 41°F (5°C).
- Mangoes are very adaptable, surviving occasional floods, droughts, and temperatures in excess of 113°F (45°C), and do not need a complex pruning regime.

Eaten fresh, the mango is one of the finest of fruits, but not necessarily one that everyone takes to at first because, for the uninitiated, the aromatic element can suggest petroleum. There are many different ways of preparing the fruit, which takes some skill because the stone is not clearly differentiated from the juicy flesh. Mango juice is also gaining in popularity.

Mango wood shows a rich array of colors and patterns, and with the global increase in mango cultivation, its wood is increasingly available for furniture manufacture. The bold grain gives individual appeal to the pieces. Mango wood is naturally light in color, with furniture finished with a clear varnish, but much is also stained to resemble dark woods, such as mahogany.

Up to 65 ft. (20 m)

Up to 130 ft. (40 m)

Mango wood and foliage contain a number of chemicals, at least one of which is the same as one of the active ingredients in poison ivy. The plant causes contact dermatitis in sensitive individuals living in countries where it is grown.

Mango was once only available in canned form outside its production zone. A more modern method of preservation involves drying small pieces of fruit, often for inclusion in luxury breakfast cereals and snacks.

Summer in India for foodies is synonymous with the mango season. In our country, each state boasts of different varieties of mangoes, all hailed as delicacies.

"MANGO, THE KING OF FRUITS IS HERE" IN *TIMES OF INDIA*, MAY 25, 2013

Melaleuca alternifolia

Tea Tree

This large shrub, or sometimes a small tree, grows to 20 feet (6 m) in height. Tea tree belongs to the aromatic myrtle family (Myrtaceae), and this species is particularly aromatic, being the main source of the essential oil known as "tea tree" or "ti tree" oil. The tree is native to southeastern Australia and was used by Aboriginal peoples for a variety of medical purposes. Its versatility and efficacy led to its being taken up widely by complementary health practitioners in the 1980s.

The tree has an upright habit, with its foliage a dense mass of narrow, almost needlelike, but soft, leaves. The flowers take the form of a dense clump of white stamens and are followed by seed pods containing a large number of very small seeds.

Tea tree oil is popular as an ingredient of shampoos and soaps, as much for its strong and healthy-smelling aroma as its undoubted medicinal properties.

The leaves, and the essential oil derived from them, are antiseptic and fungicidal. The oil has been used to treat a very wide range of skin conditions, and there is evidence that it does work in many cases. Its original Aboriginal use was as a treatment for cuts and abrasions, where its antiseptic qualities came into play.

Up to 20 ft. (6 m)

Up to 20 ft. (6 m)

⚠ Tea tree oil is toxic and therefore only suitable for external use; it should never be ingested. Even externally it can be harmful: repeated application to the skin, especially undiluted, may result in irritation and dermatitis.

- Seed germinates quickly in warm conditions, but care must be taken that the tiny seedlings do not dry out.
- Seedlings grow relatively fast, and young plants establish quickly, too.
- Full sun and moist but well-drained soil are required.
- The tree is hardy to around 23°F (-5°C).
- If the tree is grown indoors, regular pruning is needed to keep growth in check.

Melaleuca quinquenervia ⚠

Paperbark Tree

Melaleuca quinquenervia is one of the largest species of an extensive evergreen genus, growing to a maximum height of 65 feet (20 m). It is found along a coastal strip in savanna and seasonally inundated areas of Australia, as well as in New Caledonia and part of Papua New Guinea. In the southeastern United States it has become a major invasive weed, and is regarded as one of the greatest threats to the ecological integrity of the Everglades region in Florida. It has remarkable powers of recovery and can resprout from its base or trunk within weeks of a fire. The flowers are typical of *Melaleuca* and its relatives, being masses of fluffy stamens, in this case cream in color. It produces very large quantities of fine seed, and disperses these very effectively.

Aboriginal peoples in Australia used the peeling paperlike bark of the tree to make coolamons, a kind of boat-shaped container used for transporting babies, food, and water, and for making shelters and wrapping food.

The aromatic leaves are used by indigenous Australians to prepare a tea for treating colds, coughs, and other minor ailments. They are also effective as a treatment for hay fever and sinus infections, being crushed and the vapor inhaled. The aromatic oils are similar to those of eucalypts.

• Seed germinates rapidly and easily, although the seedlings are very small and easily lost.
• Young trees grow well in subtropical climates, and are able to cope with both drought and flooding.
• There is little frost resistance.
• Given the invasive potential of this species, it is inadvisable to grow it anywhere where it could spread.

Up to 35 ft. (10 m)

Up to 65 ft. (20 m)

The pleasing aroma of the tree has been captured through the distillation of certain compounds. The tree's essential oil has been used in perfumery, especially in Australia.

Bodies would be wrapped in the bark before the funeral ceremony commenced.

"AUSTRALIA'S PAPER BARK: THE MELALEUCA," ON THRIVE ON NEWS WEBSITE

Each fruit contains five seeds, which, having a natural perforation through the center, make ideal beads.

"MELIA AZEDARACH—BEAD TREE," TOP TROPICALS WEBSITE

Chinaberry

Native to northern India, Southeast Asia, and Australasia, the chinaberry normally grows to around 65 feet (20 m) in height. With pinnate leaves and attractive clusters of fragrant, lilac-colored flowers, it is ornamental enough to have been introduced to regions well beyond its homeland. It is sometimes used as a street tree, despite problems caused by its slimy fruit. This fruit is a berrylike structure that, if it does not fall off the tree, eventually dries and hardens to form a distinctive five-sided nut about the size of a marble.

Like its relative, the neem tree (*Azadirachta indica*, see p. 38), the chinaberry has tissues that contain a compound, azadirachtin, which acts as a natural insecticide, or at least deterrent to insect feeding, helping keep the plant free of pests. The compound also helps to inhibit the decay of the wood. In addition, it possibly has an inhibitory effect on the growth of species growing beneath the tree. This characteristic, along with prolific seeding and the production of suckers from the roots, means that the tree can rapidly spread into disturbed habitats.

- Soaking the seed in hot water or scarifying with sandpaper both help to reduce the time the seed takes to germinate.
- Young plants grow rapidly and the species is adaptable to a wide range of soil types.
- The young tree should be watered regularly; it will tolerate drought and needs no irrigation in winter. Chinaberry takes pruning and coppicing well.

The chinaberry is in the same family as mahogany and the timber is of high quality, dark enough to look like teak, and easily worked. Given that the tree is common and often seen as a problem weed species, it is surprising that the potential of the tree as a source of quality wood is not more realized. What wood is available, cut from local trees, tends to be used by hobbyists for veneers, carvings, and turned objects.

Regarded as an invasive alien in some areas, including the southern states of the United States, chinaberry produces fruits that are poisonous to humans when eaten in quantity. Many birds are attracted to the berries without serious ill-effect, although the toxicity is observed in their subsequent inebriated behavior. For humans a greater risk lies in the extreme slipperiness of the fruit when it falls off the tree onto road and sidewalk surfaces, giving rise to unexpected accidents and falls.

Up to 80 ft. (25 m)

Up to 65 ft. (20 m)

Dried chinaberry fruits are used for making beads, including rosary beads, for which they are particularly suitable as frequent handling brings a shiny polish to their surface. Left on the branches the berries are decorative enough to be used by florists.

The fruit and the leaves may be used as a deterrent to vermin, so long as there is no risk of their being accidentally mixed up with food. A decoction of the leaves has been used as a natural pesticide.

Moringa, Drumstick Tree

A short-lived, fast-growing tree that reaches a maximum height of 40 feet (12 m), this intriguing species has recently achieved a reputation as a "wonder tree" for developing countries. Such claims have been made for other plant species in the past, but in this case there appears to be good evidence to support its more widespread cultivation.

Found wild across a semi-arid or seasonally dry zone stretching from Afghanistan to Bangladesh, the plant, also known as the horseradish tree, has long been cultivated in similar climate zones across the world. It has leaves divided into threes, corky bark, and—unusually for a tree—flowers in its first year, to produce long pods that are eaten as a vegetable. The tree can be grown annually, or coppiced, so that pods are easily reached and the foliage cut. Its cultivation is very easy across a wide range of conditions.

In recent years greater attention has been focused on the tree because of the supposedly healthy qualities of the plant. Traditional medical systems, especially in Africa, have for a long time prescribed the tree for a variety of diseases. With the spread of AIDS across Africa, research into cost-effective and local solutions became a necessity.

- Moringa is easy and quick to grow from both seed and cuttings. Both techniques can be used directly for the final positioning of plants—a benefit when the trees are used for soil-erosion control.
- Adapted to seasonally dry climates, moringa can be grown anywhere that is frost free.
- In colder climates, it can be grown as a summer annual. It does not need fertile soils or high levels of irrigation, and takes hard pruning well.

An increasing body of scientific research, and much anecdotal evidence, supports the claim that moringa products have an antibiotic effect—for example, in the treatment of urinary tract infections, pneumonia, tuberculosis, and meningitis.

Up to 35 ft. (10 m)

Up to 40 ft. (12 m)

A popular vegetable in India, moringa "drumsticks" are seed pods that may be pulled through the teeth to extract the pulp inside; the tough skins are inedible and discarded. The leaves may also be eaten, although in many places they are only fed to livestock.

Oil made from the seeds has a wide variety of uses, including a lubricant for small-scale machinery. It can also be used in cosmetics and in cooking. The crushed kernels may also be used to remove solids from turbid water.

Moringa planted directly into the ground rapidly makes an effective erosion control or a living fence.

Claims for the use of moringa against AIDS probably owe more to its being a very good source of micronutrients, helping patient survival, than any antiviral effect it may have. In poor countries, such food sources can be enough to ensure survival from disease, as well as ensure a woman's health in pregnancy.

Moringa leaves contain more Vitamin A than carrots, more calcium than milk, more iron than spinach, more Vitamin C than oranges, and more potassium than bananas.

J. W. FAHEY, "MORINGA OLEIFERA," IN TREES FOR LIFE JOURNAL

> The white mulberry forms an excellent
> live fence, and when established is the
> most permanent of any other. Cattle must
> not be allowed free access . . . while young
> . . . but after it has become a good fence
> they may approach it with advantage.

J. A. CLARKE, *TREATISE ON THE MULBERRY TREE AND SILKWORM:*
AND ON THE PRODUCTION AND MANUFACTURE OF SILK (1832)

White Mulberry

This deciduous broad-leaved tree, growing to around 65 feet (20 m), originated in northern China but is now very widely grown in suitable climates. Although its homeland has a continental climate, with cold winters, it has proved adaptable to warmer regions. Male and female flowers are generally on different trees, with pollen from the males being blown by wind onto the females. To send the pollen out into the breeze, the male flowers have evolved a catapult mechanism that ejects the pollen at 350 miles per hour (560 km/h)—at half the speed of sound, the fastest movement in the plant kingdom. The female flowers mature to dark red raspberry-like berries.

However, it is more for the leaves that the tree is widely distributed. On mature trees they are broadly oval, up to 6 inches (15 cm) long with a pointed tip; on younger trees or young shoots they are deeply lobed and much larger. They are soft, highly nutritious, and, crucially, are the preferred food for a moth caterpillar whose chrysalis is the basis of one of the most highly prized products of the natural world—silk. Trees grown in order to feed silkworms are generally grown in rows and are frequently cut back, so that at first sight they look like grape vines.

- White mulberry seed germinates well, preferably after three months of stratification.
- Growing from cuttings is also relatively easy.
- Growth tends to be slow, and is best in warm-summer climates, with fertile, moist, but well-drained soil, or good irrigation.
- There are a number of cultivars, including at least one dwarf one.
- Be warned that it hybridizes with the red mulberry.

The root bark is used in some Asian traditional medicinal systems, and modern research shows that it does indeed have some valuable properties, primarily against bacteria. This gives it some potential for use against infections, including those related to tooth decay. The immature fruits are toxic to humans.

White mulberry trees are grown on a huge scale across the world to feed silkworms. Historically, many rulers tried to encourage silk production, but of course this was only successful where white mulberry could be cultivated. They suit smallholder agriculture because the roots tend to plunge deep down in the soil, with little in the way of surface rooting; it is therefore possible to grow other crops close to them.

Mulberry foliage is not favored only by silkworms; the leaves are equally good for feeding livestock. Their use as cattle fodder is growing across the world, especially in the warmer climates the tree prefers.

White mulberry fruit is inferior in flavor to that of the black mulberry (*Morus nigra*). Even so, where trees are grown large enough to bear fruit they are either eaten fresh or dried for later consumption.

Up to 40 ft. (12 m)

Up to 65 ft. (20 m)

Murraya koenigii

Curry Leaf

This is a small tree, growing only to 20 feet (6 m), with evergreen leaves divided into many leaflets, each one about ¾ inch (2 cm) long. With their distinct and pleasant aroma, the leaves feature in many Asian cuisines. The tree's clusters of small, white, fragrant flowers are followed by black berries. Curry leaf is a tropical species and is native to the warmer, more humid, southern half of the Indian subcontinent, but the tree is frequently planted elsewhere.

The leaf shape is very similar to that of the neem tree (*Azadirachta indica*, see p. 38), and most people perceive the aroma as similar. In many Indian languages, curry leaf is referred to as the "sweet neem," as opposed to the "bitter neem." Several different cultivars vary in their growth rate, size, and quality of leaf; "Gamthi" is the most aromatic.

- Seed germinates regularly at around 68°F (20°C), and the plant can also be propagated by cuttings.
- Being from the humid tropics, it is not tolerant of cold or drying out.
- Keeping it well fed and well pruned stimulates extra production of leaf shoots.

Hard to describe, the flavor is best understood as a background element in savory dishes where other spices have more prominence. The leaf is used in a very wide variety of south Indian or Sri Lankan dishes, and also appears in the cuisines of Southeast Asian countries, such as Cambodia and Vietnam.

Up to 13 ft. (4 m)

Up to 20 ft. (6 m)

In Ayurvedic medicine the leaf is used to treat sickness, diarrhea, and diabetes. Western researchers have confirmed that it contains compounds that inhibit the action of a pancreatic enzyme, slowing the release of glucose—a very desirable therapeutic effect.

A traditional oil for the hair is made by boiling curry leaves in coconut oil. It is massaged into the scalp to promote hair growth and delay graying. An essential oil extracted from the leaves is also used in the making of cosmetics.

Musa × paradisiaca

Banana

Technically, the banana is not a tree but a large herbaceous perennial. The plant is included here because it is popularly thought of as a tree, a confusion that should be rectified. People with no experience of the tropics find it difficult to believe that a plant capable of growing to 23 feet (7 m) is not a tree, but it forms no woody growth and each of its shoots is relatively short-lived.

Banana plants are genetically complex, the result of several different species crossing in prehistory. It is now thought that they were first domesticated in Southeast Asia, possibly as long as 8,000 years ago. They are sterile—if the fruit contained seeds they would be much harder to eat. Today, their center of diversity is southern India, where a host of different varieties exists.

Banana flowers are used in treating bronchitis, dysentery, and ulcers. Many other parts of the plant are used, too; for example, the leaves contain an enzyme that has a role in treatments for Parkinson's disease.

It is customary to serve meals on banana leaves instead of plates in southern India. Doing so makes dish washing unnecessary, and leftovers can simply be wrapped up in the leaf and the two composted together.

Up to 15 ft. (4.5 m)

Up to 23 ft. (7 m)

Bananas are a staple food for many Africans, and contribute a lot to diets elsewhere in the tropics. Varieties that are less sweet, and grown primarily for their carbohydrate, may be called plantains. Smaller varieties with distinct flavors may be associated with dishes for particular festivals.

- The only bananas that can be grown from seed are ornamental varieties that produce fertile fruit.
- Some species can be grown as ornamentals in colder climates and are hardy to around 14°F (-10°C).
- Fruiting bananas require subtropical or tropical conditions, fertile soils, and good levels of moisture.

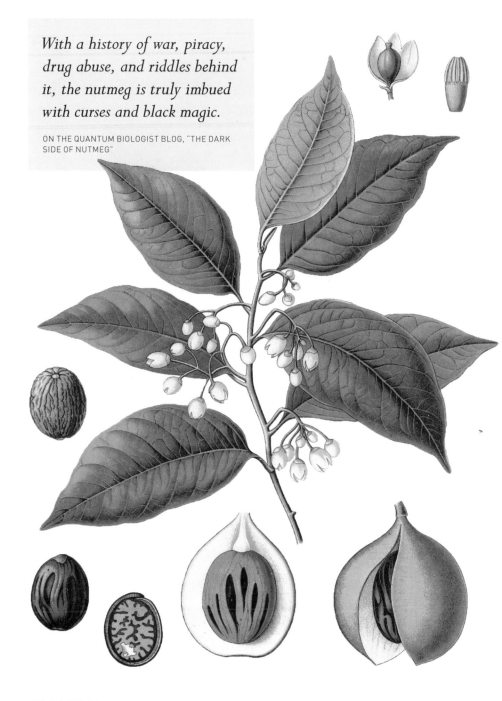

> *With a history of war, piracy, drug abuse, and riddles behind it, the nutmeg is truly imbued with curses and black magic.*
>
> ON THE QUANTUM BIOLOGIST BLOG, "THE DARK SIDE OF NUTMEG"

Myristica fragrans

Nutmeg

An evergreen tree growing to a maximum height of 65 feet (20 m), and originally native only to the Banda Islands in Indonesia, the nutmeg is now grown commercially in Southeast Asia, the Caribbean, and in southern India. Growing on the tree, the fruits look like apricots. The seed found within is the source of the spice nutmeg, while a lacy coating around the seed is the source of a second spice, called mace.

Nutmeg has always been one of the most highly regarded spices. Long known in Byzantium, and exported to Europe from the late medieval period onward, its popularity and high price helped to drive European exploration eastward. The restricted source of nutmeg, which had been a closely guarded secret of the local trading community, was discovered by the Portuguese in the sixteenth century. A century later, the Dutch gained control of the islands after a struggle with the British and their traders massacred much of the local population and enslaved the survivors as part of an attempt to achieve a monopoly of production of nutmeg, mace, and another local spice: cloves. During the early nineteenth-century Napoleonic Wars, the British succeeded in occupying the islands, after which they carried away nutmeg trees to start their own plantations in the Caribbean.

- Fresh nutmeg seeds germinate easily at 68–77°F (20–25°C), and may be grown on in constant warm and moist conditions.
- Male and female plants are distinct, and only females bear nutmegs. Most commercial production relies on grafting one male for every twenty females to ensure pollination.
- Trees begin to bear fruit from seven years onward.
- Nutmegs prefer the light shade of taller trees to standing in open sunlight.

Nutmeg and mace both have a sweet flavor that, although highly distinctive, is not powerful or overwhelming. Despite the sweetness, both are used predominantly in savory dishes. Mace is more delicate in flavor than nutmeg, and has an attractive yellow color that it can impart to food. The spices have a long history of use in a wide range of cuisines, originally in East Asia and then in the West.

Traditionally included as an ingredient in cigarettes in Indonesia, nutmeg has long been known to have mild hallucinogenic properties. Unfortunately for thrill-seekers, the dose needed to get any effect is quite close to that for toxicity, and excessive consumption usually causes severe nausea and headaches.

Nutmeg is little used in traditional medical systems, although one exception is the use of it against malaria in Thailand. However, research indicates that it may have some beneficial impact on memory, and against tooth decay.

Nutmegs can be crushed to produce oil, which is then used as a flavoring, often in conjunction with cinnamon, in products such as coffee, toothpaste, and cough syrups. In Scandinavia nutmeg is used to flavor sausages.

Up to 50 ft. (15 m)

Up to 65 ft. (20 m)

Olea europaea

Olive

Growing to a maximum height of 50 feet (15 m), the olive is a modestly sized species native to the eastern Mediterranean region. It is one of the world's most widely grown trees, and one that has had a remarkably long and intimate relationship with humanity. With its compact habit and stiff, pointed, grayish leaves, the olive is immediately recognizable in its favored dry environment. Insignificant flowers produce the familiar fruit, which are increasingly harvested and processed industrially, but traditionally were gathered by whole communities, to be made into oil and other products by local artisans.

Olives are very long lived, up to two thousand years, and many elderly trees continue to be productive. They are easy to cultivate and many trees continue to bear good crops with almost no care; they are also remarkably easy to transplant. Typically, olive plantations would be part of a complex farming system involving many different crops. Commercial planters, though, aim at creating vast plantations with combinations of different varieties, to help maximize pollination. Cultivation is thought to have begun several thousand years BCE, in what is now Jordan; central to Greek and Roman civilization, the tree and knowledge of its cultivation was widely spread in Classical times.

- The pits (or seeds) of store-bought olives have been soaked in brine and will not germinate.
- Commercially, the trees are propagated by grafting shoots of known varieties onto rootstocks of wild olive seedlings.
- The trees thrive in a Mediterranean climate, but need some winter cold to stimulate flowering; they survive frosts down to around 18°F (-8°C).
- Productivity depends on skilled pruning.

Soap has long been made from olive oil mixed with soda ash or lye, but the neat oil has a long history as a skin treatment. It is particularly popular for massaging babies.

Olive oil is produced by pressing the fruit. Central to the diet of the Mediterranean and Middle Eastern regions, the oil is now used in the manufacture of a wide range of food products, including margarine. It has a reputation for being a "healthy" oil, although some claims made for it should be regarded as questionable. The highest quality is "extra virgin" and cold-pressed, a process that causes the minimum of degradation of flavor.

Eating the fruit itself, rather than processing it into oil, is a growing global trend. Olives are unpalatable unless immersed in brine, and then partially fermented. The flavor of the final product depends as much on the treatment process as on the variety of olive being treated.

Olive trees do not produce much timber, so the very hard, durable, and attractively patterned wood is only used for small, relatively high-value items.

Up to 40 ft. (12 m)

Up to 50 ft. (15 m)

For the Lord thy God bringeth thee into a good land . . . a land of oil, olive, and honey.

FROM THE BIBLE,
DEUTERONOMY 8:7 AND 8:8

> The Paulownia tree has been grown in China for at least 2,600 years. It may well hold the record for history's oldest plantation tree.
>
> TIMOTHY HALL, "PAULOWNIA: AN AGROFORESTRY GEM"

Paulownia, Foxglove Tree

A fast-growing deciduous pioneer tree species, potentially growing to a height of 80 feet (25 m), the paulownia is native to western China. It was carried to Japan millennia ago and there is a long history of its being cultivated there; in fact, a stylized image of the tree serves as the official emblem of the Prime Minister's Office. The tree is not tolerant of shade so is eventually displaced by other species. However, it is able to persist in areas where wildfires are common, as it can sprout again from the ground if the top growth is burned off. The same happens if the top is killed by a cold winter.

The leaves are typically 8 to 16 inches (20–40 cm) across, but trees that are coppiced can produce new growth with leaves twice that size. The exotic and dramatic foliage can feature strongly in ornamental plantings. The flowers are trumpet-shaped, purple, and grow in bunches. The fact that they appear before the leaves adds to the ornamental value of the tree. The seed is winged and very fine, and the tree produces it in large quantities, a habit typical of pioneer species. The paulownia's ability to spread and dominate disturbed habitats has caused it to have a reputation as an invasive alien in some parts of the United States.

- Seed germinates readily within two months.
- The young plants make rapid progress in climates with warm summers. As long as water is plentiful, they can make several feet of growth a year and flower in five years.
- In cooler climates, growth will be much slower.
- Although the tree is hardy to around 14°F (-10°C), frosts can damage flower buds.
- There is little tolerance of drought.

Paulownia species are becoming increasingly popular in agroforestry, although plantations are only practicable in regions with sufficient water to maintain a fast growth rate. The wood is light, soft, and warp-resistant, but is not regarded as being of particularly high quality. It does make good charcoal, however.

The wood has long been popular in East Asia for making furniture and the soundboards of stringed musical instruments such as the guqin and the kayagum. A tradition recorded in China and Japan is that a tree is planted on the birth of a baby girl; when she marries, the tree is felled to make a cabinet for her dowry.

Paulownia makes a good bio-remediation tree. Waste from animal husbandry can be piped into plantations, where the trees' physiology ensures the rapid absorption of nutrients. This action helps to prevent the pollution of water courses.

The seeds are so light and plentiful that in the early twentieth century Chinese immigrants to the United States used them to pack delicate items. Leakage of seed from packing boxes is thought to have spread the tree along railroad lines.

Up to 60 ft. (18 m)

Up to 80 ft. (25 m)

Avocado

The avocado is an evergreen tree capable of reaching a height of 65 feet (20 m), and its high-calorie fruit is an important energy food. It originated in central Mexico, and archaeological evidence suggests that it has played a part in the complex and efficient agricultural systems of that region for thousands of years. Since its discovery by Europeans in the sixteenth century, it has been widely cultivated in warm-climate regions around the world, despite being somewhat demanding in its requirements—the sites for avocado orchards must be selected carefully. Mexico remains the largest producer, with other regions of importance including California, Israel, and Southeast Asia.

Avocado flowers avoid self-pollination by opening either as male the first day and female the next, or the other way round. The order of opening varies between cultivars. Like many fruits, it does not guarantee a good harvest every year; many cultivars will bear significantly only in alternate years.

Some 80 percent of global fruit production derives from the single cultivar "Hass," which originated in California in the 1930s. This produces high-quality fruit all year round, and, very importantly, is able to survive cold snaps better than most other alternative cultivars.

- Avocado seeds germinate readily if placed in warm, moist soil.
- Avocados are frequently grown as houseplants.
- To flourish outside, they must have warm conditions and moist soil all year, with at least 100 cm (40 in.) of rainfall per year.
- Very light overnight frosts are tolerated.
- Cultivars are generally propagated by grafting.
- Cultivars vary considerably in the climatic conditions they prefer.

Avocados have the highest energy content of any fruit, largely because of their fat content—in some regions they are known as "poor man's butter." While often eaten unprepared, or with an oil-and-vinegar dressing, they are used in a wide range of recipes, the most well-known being the Mexican guacamole, where the fruit is mashed, and then blended with chili and other flavorings. Both savory and sweet avocado recipes are found across a wide range of cooking cultures.

Leaf extracts show anti-fungal and antibacterial properties as well as promoting root growth. These characteristics make the leaves potentially useful for making a dip for cuttings in plant propagation.

Up to 35 ft. (10 m)

Up to 65 ft. (20 m)

The oil produced by pressing the fruit is absorbed rapidly by the skin. This has led to its extensive use by the cosmetics industry as an ingredient.

Avocado leaves are toxic to mammals and birds, and also to insects, to the extent that they may be used to make an insecticide. Despite this potential toxicity, some herbal practitioners promote a tea made with avocado leaf for lowering blood pressure and for cleansing the kidneys by flushing out unwanted waste.

The Avocado or Alligator Pear-Tree. It grows in gardens and fields throughout Jamaica.

THE FIRST PUBLISHED USE OF THE NAME "AVOCADO,"
IN A CATALOGUE OF JAMAICAN PLANTS PUBLISHED
BY NATURALIST SIR HANS SLOAN IN 1696

Phoenix dactylifera

Date Palm

Growing to 65 feet (20 m) high, the date palm is possibly one of the oldest plant species in cultivation. Its relationship with humanity was particularly intimate in the case of the desert nomads and traders who appreciated its fruit and eventually carried the tree from its original home in the valley of the Euphrates and Tigris (in present-day Iraq), spreading it widely in oases from Morocco across to India. This spread was facilitated by their shared adherence to and unification under Islam. In the nineteenth century, the tree was taken still further by European adventurers to the far reaches of Western empires. It is now an important crop in the United States and China as well as the Middle East.

The trees have male and female flowers on separate plants, a factor that was realized by the ancient civilizations of the Middle East. This was the first example of humanity understanding sexual reproduction in plants, and the necessity of growing the right proportions of male to female plants in plantations. Distinct varieties were soon picked out and propagated by suckers. Continued development remains important in the Middle East; in 2009, a team in Qatar unraveled the genome of the date, opening up new ways to improve the tree as a crop.

- Date pits will germinate in a few weeks at around 68°F (20°C), after which they will grow rapidly.
- They should be potted on and grown in containers until planting out in their second year.
- Date palms need full sunlight and fertile soil; they are oasis plants, not desert ones, and will not tolerate regular prolonged drought.
- They can withstand occasional overnight frosts down to 19°F (-7°C).

The fruit is used in a wide range of dishes, predominantly sweet ones, and is often added as a paste or syrup. From the eighteenth century onward its ease of transportation led to preserved dates being used in an increasing number of Western dishes, too. Fresh dates have always been popular, particularly those of the sweeter and juicier varieties.

The leaves, or parts of them, have many more traditional uses than the low-quality wood, being used in waterproof roofing for huts, baskets, mats, and even rope. A modern use is in the making of cheap insulation board, which not only keeps the occupants of basic homes warm but also reduces energy needs, which has a positive impact on the environment.

Up to 30 ft. (9 m)

Up to 65 ft. (20 m)

Dates are packed with sugars and nutrients, and can be dried, in which condition they will keep for years. They are ideal as easily carried emergency rations, hence their high status with the desert peoples of Africa and Asia. In some cases, the dates are pitted for storage as compacted blocks.

Date palms can be tapped in the way of rubber trees for their sap, which is used either to produce palm wine or to be boiled down into "jaggery" sugar. Tapping risks killing the tree, and needs considerable skill.

And from the fruits of date-palms and grapes, you derive strong drink and a goodly provision. Verily, therein is indeed a sign for people who have wisdom.

KORAN, "AN-NAHL," CHAPTER 16, VERSE 67, MOHSIN KHAN TRANSLATION

Named an Oregon Heritage Tree in 1997, the Klootchy Creek Sitka sprouted a few decades after the Magna Carta was signed in 1215.

KIM POKORNY, "OREGON HERITAGE TREE: KLOOTCHY CREEK SITKA SPRUCE"

Picea sitchensis

Sitka Spruce

With the tallest specimens growing just short of 330 feet (100m), Sitka spruce is one of the world's big trees, feeding on the plentiful rainfall of a belt of land along the northwestern coast of North America. It is the third tallest conifer species. Logging started early after the arrival of European settlers and the tallest specimens were probably lost in the nineteenth century; old-growth survivors are now protected.

Today, the species is widely planted for forestry, because it grows rapidly in high-rainfall regions such as the west of Britain, western Norway, and New Zealand, where it flourishes even on poor soils and exposed sites. The trees' very fast growth rate makes it economically viable to harvest them in as little as forty years from planting—in fact, in many temperate regions it is simply the most productive tree that can be grown. However, these forestry plantations are extremely dense, and are widely unpopular with ecologists and the general public for their lack of diversity and their gloomy impact on the landscape.

The needles look like those of any other spruce but are notably sharp, which has precluded its use as Christmas trees. The cones are pendant, up to 4 inches (10cm) long, and have distinctly prominent bracts.

- The seed germinates easily, although winter stratification is usual in commercial production.
- Seedlings are grown on in pots for a year before planting out.
- Growth is rapid in good conditions, namely high rainfall and acidic but mineral-rich soils.
- There is little tolerance of drought, or waterlogging, and the species does not regenerate well after cutting or pruning.

Sitka spruce is remarkable for the many long, mainly undivided roots that it sends out just below the surface in all directions, often extending beyond the width of the canopy. Roots may be more than 50 feet (15m) long. Native Americans (particularly Alaskan tribes) would dig these up, strip off their bark, and use the strong inner fibers to weave beautiful baskets and other items. The roots are strong and flexible, and Native Americans would also split them for various uses, including twine for sewing, fishing line, and string.

The timber tends to be straight with few knots, and is very strong for its weight. The Wright Brothers used it in their pioneering aircraft, and its use in aviation continued for some decades. The wood conducts sound extremely well and is therefore used in the making of musical instruments, notably the soundboards of acoustic guitars. Otherwise, its use is very much an industrial one, primarily as a major source of cheap but strong wood for the construction industry, and as material for pulping.

The tips of Sitka spruce twigs are one of the best sources of vitamin C in its region, and Native Americans would eat them in the winter months to ward off scurvy. The tips may be eaten as a vegetable, brewed into a tea with a light, citrus flavor, or used in the making of spruce beer. In the spring, the sap of Sitka spruce can be tapped and drunk, too.

Up to 80ft. (25 m)

Up to 330 ft. (100 m)

Longleaf Pine

Few trees so dominated their region and habitat as the longleaf pine did in the American South before the arrival of the European settlers. Growing to 115 feet (35 m), *Pinus palustris* formed forests that stretched for hundreds of miles, with the floor dominated by a complex and diverse plant community. Every decade or so, fire would sweep through, but only on the ground, doing little damage to adult trees. Fire was a vital part of the ecology of the region, ensuring the long-term dominance of longleaf and other species that were either fire-resistant or could re-establish themselves. Longleaf seedlings keep their vulnerable growth point buried deep in a grasslike cluster of needles.

Longleaf—so-called because of its needle length of up to 18 inches (45 cm)—is now something of an icon for conservation organizations in the South. Only a tiny fraction of virgin forest remains, and conservation and ecology groups have long sought to protect those trees. A growing realization that the species is not only crucial to a healthy ecosystem but also commercially valuable has brought landowners, foresters, and the military (who use some of the finest remaining stands as training grounds) into a coalition to replant it.

- Seeds require around two months of cold stratification before being sown at around 68°F (20°C); germination tends to start from two weeks after that.
- Seedlings should be grown on in containers for a year or two before planting out, in full sun, with care being taken not to plant too deep.
- The seedlings will tolerate infertile and acidic soils, but will do best with good drainage and plentiful moisture.

Longleaf pine produces timber of high quality, which accounts for the alacrity with which it was felled during the nineteenth century, when large amounts were used in construction and for furnishing. Boards from this period could be exceptionally wide, to 3 feet (1 m); they are now much sought after as architectural salvage.

Native Americans would use soaked longleaf needles to make baskets. They tied bunches of the needles together with twine, then coiled the flexible lengths that resulted up to the desired height, in the way of ceramic coil pots. The bottom of the basket is made from a flat coil. Craft enthusiasts have kept the tradition alive.

Up to 35 ft. (10 m)

Up to 115 ft. (35 m)

Old longleaf pine stumps, saturated with resin, are still being dug up from fields, long after the trees were felled. The wood is sought after for fire-lighting—it is highly flammable, even after many decades of exposure to the weather.

Heated in a kiln, longleaf pine branches and timber offcuts produce a thick tar, derived from the concentrated resin. In the past the tar was used to weatherproof rigging and other fittings on sailing ships. Longleaf produced some of the best tar, and its production was once a major industry. A concentrated form, known as "pitch," was also used for sealing and waterproofing timbers, on houses as well as ships.

Starting in southwest
Virginia, the longleaf
pine forest stretched southward
through nine states, eventually
stopping in east Texas
(over 140,000 square miles).

"OVERVIEW—THE BIG PICTURE,"
ON LONGLEAF ALLIANCE WEBSITE

Chir Pine

Chir pine is the characteristic pine of the Himalayan foothills of the northern Indian subcontinent. The impressively large cones (up to 10 x 3 in. /25 x 8 cm) open gradually or during a fire—like many pines, this is a species that has evolved as a pioneer colonizer after forest fires. The thick bark of the tree helps protect mature specimens from the effects of ground fire. The other major tree of the region is *Quercus leucotrichophora* (banj oak), which is a longer-lived dominant. During the late nineteenth and early twentieth centuries, British colonial forestry authorities encouraged the felling of banj, and its replacement with chir because the latter was proving a useful source of industrial products. This led to much local resentment and during the Chipko movement of the 1970s, local people, especially women, led a campaign to save the forests of the Uttarakhand region from being felled by outsiders. Chir needs more water than banj, and in drier areas the spread of chir has noticeably reduced the water available to the human population.

- Seed needs several weeks of cold stratification before sowing. Seedlings develop fast.
- Plants need a well-drained soil (preferably dry or moist) in full sun.
- They can tolerate drought but not much frost.

The local people collect the long needles of the tree for use as animal bedding. They also carve the bark of the tree into lids for vessels.

Chir became important during the time of the British Raj in India when it was discovered that the resin could be distilled to produce turpentine. In an era before petrochemicals, this was the start of a series of chemical pathways leading to products used in making rubber, pharmaceuticals, cosmetics and a host of other applications.

Up to 4 ft. (120 cm)

Up to 165 ft. (50 m)

Named after William Roxburgh ... the Father of Indian Botany.

HILLS OF MORNI WEBSITE

Pinus sylvestris

Scots Pine

Found from the western fringes of Europe to the edge of the Pacific, the Scots pine is one of the most widely distributed of all conifers. It is the only pine native to northern Europe. Growing up to 200 feet (60 m), as the tree ages, it develops a very gappy asymmetric habit and therefore rather lacks bulk in its outline. Over most of its range, it is a climax species, dominating the landscape, sometimes in conjunction with birch. Its forest tends to be open, with a rich diversity of ground flora. One of the most famous areas of the species is the Caledonian forest of Scotland, which is now much diminished after centuries of abuse from both locals and English colonizers.

Quick-growing, durable, and easy to work, Scots Pine is the material of choice for light-colored "pine" furniture. Traditionally, however, it was used for boat building because its high content of resin helped keep the wood from decaying.

The resin has long been used to make turpentine, but the material remaining is known as "rosin," which is also known as "Greek pitch." Although today this has many industrial uses, traditionally it was used to help increase grip, for example, on dancers' shoes, violinists' bows, and gymnasts' hands.

Up to 40 ft. (12 m)

Up to 115 ft. (35 m)

The antiseptic resin or young leaves have largely been used for treating respiratory problems. Native Americans used it to treat lung infections. Rheumatism and arthritis have also been treated with a variety of products extracted from the tree.

• Seed is best sown in fall and left outside over winter, or given a month of cold stratification.
• It then germinates readily and grows quickly.
• Young trees are best kept in containers or seedbeds until they are two years old when they can be planted out.
• Good drainage and full sunshine are important, but they are tolerant of poor and acidic soils.

Pistacia lentiscus

Mastic

A shrub or small tree growing to a maximum of 23 feet (7 m), with evergreen foliage, the mastic is native to the Mediterranean, but is often introduced elsewhere. It is a pioneer plant, but also a tough survivor of environments degraded by overgrazing and deforestation; it is also frequently found as part of plant communities dominated by other species.

Mastic has a strong resinous smell and if the stems are cut, drops of aromatic resin ooze out. This material (also known as Arabic gum) has been collected since ancient times and used in various ways. Although mastic is produced in many places, only the Chios region of Greece has been granted "protected geographical indication" status by the European Union. Production is controlled by a cooperative run by local villages.

Mastic resin has a long history of food use in the eastern Mediterranean, but its piney refreshing flavor has not traveled much beyond Middle Eastern and Greek cuisine. It is used in the confection Turkish delight, which is the product those outside the region would be most familiar with.

The resin has generally been used for treating digestive problems. Recent research has shown it to have strongly antibacterial, antiseptic, and antifungal properties, with a particular potential for treating

Mastic resin has often been used as a background flavor or fragrance in cosmetic preparation. It has also been used in soaps, anti-wrinkle creams, and as a filling in dentistry.

• Seed needs two months cold stratification or fall sowing before germination will occur, which will be patchy.
• It can also be propagated from cuttings.
• It is hardy to around 14°F (-10°C), but will not cope with prolonged or damp cold.
• Good drainage is necessary, but drought, or exposure to salty or hot winds, is tolerated well.

Up to 10 ft. (3 m)

Up to 23 ft. (7 m)

Pistacia vera

Pistachio

A deciduous tree growing up to 35 feet (10 m), the pistachio is the source of one of the world's most highly prized nuts. It is a member of the cashew family. Naturally a plant of harsh desert environments, and found from the eastern Mediterranean across to Kyrgyzstan, it responds well to cultivation and has long been grown. Male and female flowers are found on separate plants, and commercially they are planted at a male to female ratio of no more than 1:12.

Archaeological remains found in the Middle East indicate that the nuts have been collected by humanity for tens of thousands of years; they were mentioned by writers in ancient Rome and were widely distributed by Islamic empires. Pistachio trees were said to have grown in the Hanging Gardens of Babylon (modern-day Hillah, Iraq) in about 700 BCE.

- Supermarket nuts are roasted and so are no longer alive.
- If fresh nuts can be obtained, they should be shelled and planted in compost at around 68°F (20°C); germination will occur from one month onward.
- The plants need full sun and grow best in seasonally dry continental climates; they will take overnight frosts down to 14°F (-10°C), but only in dry conditions.

The act of shelling pistachios in order to eat them one by one helps to reduce appetite, by making the eater feel fuller than they really are. There is evidence that they help to reduce cholesterol, although they also have a high fat content and are often salted.

Like other members of the Anacardiaceae (cashew family), the plants contain irritant chemicals that can cause allergic reactions in some people. The nuts present no danger, although they can become infected with fungus in humid growing seasons, when there is a risk of highly toxic aflatoxin contamination.

Up to 30 ft. (9 m)

Up to 35 ft. (10 m)

Pistachios are used in both savory and sweet dishes, but mostly the latter. They are a key ingredient in Middle Eastern sweet cuisine, such as baklava.

Balsam Poplar

The smell of balsam poplar in the spring is often the first sign that one of these erect-branched trees is anywhere nearby. There are few trees that are as fragrant and it is a surprisingly aromatic smell for the foliage of a cool-climate tree. The aroma is created by a resin that partially evaporates as the leaves emerge. The species is found across Canada and the far northeast of the United States; it grows up to 100 feet (30 m) and is one of the hardiest of deciduous trees. It has contributed to the gene pool known as "hybrid poplar"—a group of trees familiar from the industrial-scale plantations that are increasingly seen in temperate-zone countries, due to their very rapid growth, up to 5 feet (1.5 m) per year. They are, however, short-lived, and often inclined to be disease-prone. Balsam poplar has a long history of medicinal use and it continues to be valued in modern herbalism as an expectorant and an antiseptic tonic.

- Like all poplar seed, viability can be only a few days, and in any case, seed is frequently hybridized.
- Propagation is usually by suckers or cuttings.
- Trees grow vigorously, but need full sun and moisture.
- Tolerant of extreme cold and occasional waterlogging, but not drought.
- Felling results in the production of suckers, unless the stump is poisoned.

The resin from the leaf buds has been widely used for a variety of folk remedies—for treating sprains, wounds, chest infections, and rheumatism, among other ailments. It has also been shown to be antiseptic. The bark has been used for its anti-inflammatory and fever-lowering properties.

Balsam poplar and its hybrids are grown for the production of light, soft, but relatively durable, wood. It is easy to work and takes paint particularly well, in addition to nails and glues, making it popular with amateur woodworkers. Commercially, it is used chiefly for plywood and pulping.

Is this the balsam that the usuring senate pours into captains' wounds?

WILLIAM SHAKESPEARE, *TIMON OF ATHENS*, ACT III, SCENE 5.

Up to 25 ft. (8 m)

Up to 100 ft. (30 m)

Populus tremuloides

Quaking Aspen

Aspen is one of the most distinctive trees of the northern and mountainous areas of western North America, especially in fall when its clear pale yellow is lit up over vast areas by the sun. "Vast areas" rather describes the way the narrow tree grows—it is a clonal species, very rarely growing from seed, but instead sending out suckers from a massive interlocking root system. Aspen colonies are so successful at dominating their environment that there is no room for seedlings. Botanists think that aspens seeded at the end of the last ice age, around 10,000 years ago, and in many places their colonies have been spreading ever since. A severe winter or fire may kill trees, but they will always resprout from the base. Some groves are even older—one in Utah, dubbed "Pando," covers 106 acres (43 ha) and is possibly 80,000 years old. The "quaking" that is referred to in the common name is due to the flexible petioles of the leaves.

Aspen wood is poor quality, and is generally used only for low-quality products, such as palettes, pulping (books and newspapers), or cardboard manufacture. Pioneers of the North American west sometimes used it to build log cabins. The Aspen makes poor fuel wood as it dries slowly and decays quickly. That it does not splinter is one of its few outstanding points.

Native Americans used to cut up the bark into strips and ground it into flour, to eat as a carbohydrate source. The catkins were also eaten. The bark contains a substance that was extracted as a quinine substitute.

- Seed is very short-lived and will only germinate in a constantly moist environment, after winter chilling.
- Cuttings are the usual method of propagation.
- The trees need full sunlight, are very tolerant of extreme cold and poor soils, but need moisture and preferably good drainage.
- Once established, they start to sucker—if this is undesirable, growing in mown grass is a good way of stopping their growth.

Up to 25 ft. (8 m)

Up to 75 ft. (23 m)

And the wind full of wantonness, woos like a lover, the young aspen trees till they tremble all over.

THOMAS MOORE, *LALLA ROOKH* (1817)

The succulent, deep orange, dried apricots from the Hunza valley in northwest Pakistan have become highly respected by consumers many thousands of miles from the bazaars where they were traditionally traded.

"HUNZA APRICOTS: REACHING GREAT HEIGHTS," ON NEW AGRICULTURIST WEBSITE, SEPTEMBER, 2003.

Apricot

A deciduous fruit-bearing tree of great antiquity, apricot is found naturally across the Caucasus, Iran, and central Asia. The specific name *Prunus armeniaca* refers to Armenia, from where it was once believed to have originated, although evidence for this is slim. A tree of central Asian origin, the ancient Silk Road would have carried it and other crop species great distances, to China at one end and the Mediterranean at the other, confusing our knowledge of where they might have originally come from. Alexander the Great is said to have introduced them to Greece, from where they spread further to become one of the many fruits cultivated around the Mediterranean region.

Growing to 40 feet (12 m) high, this is somewhat larger than many fruit trees. While modern cultivation relies on small trees, each one a distinct cultivar, some traditional practices used "family trees" whereby several different ones are grafted onto one rootstock, allowing for much greater resilience in production. Although always popular fresh, throughout most of history the overwhelming bulk of the crop would have been dried for later use or trading.

All the various fruit trees that originated in central Asia flourish in an extreme continental climate—the apricot more than most, which explains it being a survival food of people in remote communities, who live in some of the harshest environments.

- Apricot pits will germinate after two months cold stratification, but there is no guarantee that when the trees fruit, they will either resemble their parents, or be any good.
- Commercially, shoots of good cultivars are grafted onto rootstocks.
- Apricots need a continental climate with cold winters, warm summers and clearly defined springs.
- Areas with late spring frosts are unsuitable for fruit production. Fertile deep soil and moisture in the growing season are important.

Apricots have traditionally been dried and either eaten later or exported to regions where they cannot be grown. Although in Western cuisines they are used almost exclusively as a dessert ingredient, in Middle Eastern cuisines they are often included in savory dishes, such as meat tagines (stews).

The kernels of apricots were often used as a source of oil by the communities that grew them; they used it for cooking, lighting lamps, or on their hair. Apricot oil is still used in cosmetics today. When cold-pressed, the kernels produce a light, gently fragrant oil that is full of fatty acids and antioxidant vitamins.

Apricot wood is seen as one of the best for smoking foods due to its mild and sweet flavor. It is good for poultry and pork.

In remote mountain areas, such as Pakistan's Hunza Valley, apricots once constituted the main food crop. Dried apricots would be made into a stew that would form the basis of many meals, with barley and any available meat.

Up to 20 ft. (6 m)

Up to 40 ft. (12 m)

Prunus avium

Sweet Cherry

Fruit from several different *Prunus* species can legitimately be called "cherries." The largest part of the gene pool of the modern cherry, however, is this species, which is a native of Europe and parts of North Africa and the Middle East. A deciduous species growing to 100 feet (30 m), but usually much shorter, this is a relatively short-lived pioneer tree, which is immediately recognizable in spring by its cream flowers and at other times of year by its distinctive shiny bark with horizontal markings. Wild cherries are edible and sometimes good to eat, but they are highly variable.

The origins of the cultivated cherry are lost in time, and it is highly probable that it was domesticated several times, as especially sweet varieties were selected from wild stock and propagated by grafting, which was invented by the Romans, who introduced the trees from Turkey. Over time, crosses were made with other species, most notably the sour or morello cherry, *Prunus cerasus*. This in turn is thought to be a hybrid between *P. avium* and another species, the shrubby *P. fruticosa*. There are a large number of varieties now available, with many being bred in the United States, for growing in specific regional climates, often for export. However, there has been relatively little success in using wild species in breeding.

- Cherry pits will germinate following a cold winter or cold stratification, but commercial varieties are always grafted.
- The trees need sunlight and deep fertile soils, moist but well-drained.
- Being relatively late flowering, they are rarely bothered by late frosts.
- The main problem is birds eating the fruit. Given the size of the tree, commercial and garden plants are generally grown on dwarfing rootstocks and kept pruned to make picking easier.

Cherries are a "fun" fruit, partly because of their small size, but also because of their very short season, and their unsuitability for drying. Nevertheless their long history and ease of cultivation has led to a number of processes and recipes being developed to make use of them. Of these, the most important has been in the making of jam. For this, and for many culinary uses, the darker varieties, known as "black cherries" have the deepest and best flavor.

Cherry wood is rich in color; it is a brown hardwood with a hint of pink or dark red that darkens with age. It is traditionally used for turning, furniture making, and for musical instruments.

Up to 50 ft. (15 m)

Up to 100 ft. (30 m)

Even without human intervention, wounds in the trunk bleed resin. This has been used medicinally, for example as an inhalant for treating persistent coughs.

Many *Prunus* species contain chemical compounds with cyanide, which can produce highly toxic hydrogen cyanide during digestion. This accounts for the bitter taste of cherry leaves and at times the fruit. Especially bitter fruit should not be eaten, but poisoning is exceedingly rare.

Cherry ripe, cherry ripe,
Ripe I cry,
Full and fair ones
Come and buy.
Cherry ripe, cherry ripe ...

ROBERT HERRICK, "CHERRY RIPE" POEM, LATER
A POPULAR SONG, c.1650.

Burbank states that his importation of twelve plum seedlings in the year 1885 was the most important importation of fruit bearers ever made at a single time into America.

PATRICK MALCOLM, "ORIGINS AND HISTORY OF PLUM TREES," ON MATRIX OF MNEMOSYNE WEBSITE.

Prunus domestica

Plum

The historic European plum is almost certainly derived from *Prunus cerasifera*, found wild across Europe and western Asia. From the nineteenth century onward, several wild species were crossed with it so that modern varieties reflect a complex gene pool. Much of the cross-breeding has been to increase hardiness or disease-resistance. Some of the most innovative work was done by Luther Burbank in California in the early twentieth century, using Asian origin species.

Plums blossom in early to mid spring, with trees smothered in small white flowers; fruit follows by early or midsummer. Their season is longer than that of cherries, but still relatively limited. Commercial growers grow several varieties, partly to overcome the issue that bedevils many fruit growers of bearing in alternate years, but also to lengthen the season, as there are several months between the earliest—and the latest—bearing cultivars. The fact that the trees have a commercial lifespan of 15 to 20 years means that turnover of orchards can be rapid. Today, plums are exported across the world as fresh fruit, but traditionally most would have been processed or preserved. Modern breeding has focused on large fruit size and tolerance of handling. Many heirloom varieties, however, are still worth growing.

- Plum pits can be germinated, but the result will be of unknown quality.
- Grafting is the standard commercial method of propagation.
- The trees are easy to grow, and being vigorous and prolifically flowering, do not need elaborate pruning programs.
- Deep fertile soils are best, with some varieties flourishing well on ones that are heavy.
- Late spring frosts can sometimes reduce cropping.

Plums are fermented and distilled to make a spirit across mainland Europe, with plum brandy (often known as slivovitz) being particularly popular in eastern Europe, where large quantities are drunk. The kernel of the pit can also be made into a spirit, with the flavor of dry amaretto.

Plum jam was a traditional favorite, but it is not as tasty as other fruit jams, so today plums are used mainly to bulk up cheaper jam blends. Different cultures preserve plums in different ways, with salting and drying being particularly common in eastern Asia. In Mexico, dried, salted plums (*saladitos*) are eaten as candy.

Dried plums are known as "prunes" and have long been used medicinally as a gentle laxative. Trying to distance their product from this usage, some growers have taken to renaming their product as "dried plums."

A number of dyes can be obtained from various parts of the plum tree, with the most important being a blue-black dye from the very dark "damson" varieties; in the past this was used for dyeing naval uniforms. A green dye is made from the leaves of some *Prunus* species.

Up to 30 ft. (9 m)

Up to 25 ft. (8 m)

Ume Plum

A deciduous tree that grows to 35 feet (10 m), it is often called "plum blossom" in English, although it is more closely related to apricot. Flowering at the very end of winter in its native Far East, it has long been a favorite tree in Chinese, Korean, and Japanese culture, where it symbolizes elegance and fortitude. For the Chinese, it is one of the "Three Friends of Winter"—plants that do not wither in the cold, and so illustrate the qualities of fortitude, perseverance, and resilience, all key virtues for the Confucian gentleman.

The species has been in cultivation for several millennia and more than 300 varieties have been selected, with flower colors ranging from white and pink to red, and doubles as well as singles. Unlike many *Prunus*, this species can live a very long time, with at least one in China reputed to be more than a thousand years old. The flowers, which are somehow perfectly proportioned with their colors emphasized by being on bare branches, have inspired countless generations of poets and artists. The tree is grown as an ornamental in North America, but suffers from too many fungal problems to make it an attractive proposition in northern Europe.

- Pits from fresh fruit need winter sowing or stratification for three months, before germinating in spring.
- Cuttings or grafting are used to propagate named cultivars.
- Young trees in fertile soil and in full sun grow quickly and flower at relatively young ages.
- The tree is cultivated for its fruit and flowers, which have a strong scent. Fruit ripens around June/July in Asia.

Plum juice is popular throughout East Asia, but is prepared using different methods by the cultures of that region. Perhaps the most distinctive is the traditional Chinese technique of smoking the plums, then boiling them, before adding sugar and fragrant herbs to the resulting juice. In Japan and Korea, a sweet and tangy juice is made from fresh green plums; the resulting refreshing drink is usually consumed in the summer. In Korea a plum syrup is made, then preserved to be drunk later after dilution with water.

Ume are used to prepare a variety of alcoholic drinks throughout East Asia, usually with a sweet flavor and sometimes blended with tea. Even people who tend not to like alcohol can rarely resist drinking Umeshu.

Up to 20 ft. (6 m)

Up to 35 ft. (10 m)

Prunus ume has long been used in traditional Chinese medicine. Recent research supports ume as having special qualities—an antiseptic action may help defend against the oral bacteria that can cause dental decay and exacerbate gastric ulcers.

Spring air

Woven moon

And plum scent.

MATSUO BASHO, HAIKU, LATE 1600s.

Black Cherry

This North American species is not to be confused with the black cherry varieties of the domestic sweet cherry (*Prunus avium*, see p. 156). It is a wild species of central and eastern North America, and parts of Central America, which grows to 80 feet (25 m), has distinctively rugged bark, spikes of small white flower and in late summer small purple-black fruit, with a bittersweet flavor. The dark gray bark of the tree emits an pepperlike smell when scratched. The tree is also known as rum cherry because early North American settlers used the fruit to flavor rum.

A pioneer species, it spreads rapidly across abandoned fields or woodland clearings, but unlike many pioneers is not particularly short-lived, although it suffers from a range of diseases, which can reduce lifespan. The fruit is eaten by wildlife, and seedlings can germinate prolifically; root damage can result in suckers being produced.

• The pits need a cold stratification period of four months before germination will occur.

• Young trees grow strongly, on any reasonable soil, but preferring deep and fertile ones.

• Seed production begins around ten years of age.

• Once trees start to fruit, seedlings can appear rapidly; in parts of central Europe it has become an invasive alien.

The leaves contain cyanogenic glycosides, which can result in illness (although this is rarely fatal) in cattle, but since the tree is so common, it is difficult for farmers to protect their livestock or eliminate the trees. The flesh of the fruit is safe to eat.

Cherry is one of the most highly prized North American timbers, being rich red-brown in color, with an attractive luster; it is relatively strong, easy to work and does not warp readily. It is much in demand for making flooring, furniture, and musical and scientific instruments. The wood is also used for smoking foods.

The fruit is edible and adds a pleasurably sharp touch to any recipe requiring cherries. It is often paired with dark chocolate in cakes.

Up to 35 ft. (10 m)

Up to 80 ft. (25 m)

Blackthorn, Sloe

Growing to only 15 feet (5 m), and often not with a clear trunk, the blackthorn or sloe verges on being a shrub. Its habit is often obscured by the fact that it suckers vigorously, forming extensive areas of scrub. These are practically impenetrable owing to both the dense branching habit and the very long thorns; these are capable of penetrating car tires. Small white flowers are borne in spring, before the leaves open, ripening to fruits, which look like miniature plums, but with an extremely astringent flavor. The species can be found all over Europe, into western Asia and the northwestern corner of Africa. This deciduous tree plays an important ecological role, as its foliage is eaten by a variety of Lepidoptera larvae, and its dense spiny branches provide a safe roosting and nesting habitat for birds.

- Fall sowing or three months stratification are needed, even then seed may take eighteen months to germinate.
- Plants will tolerate a wide range of conditions, including severe exposure and coastal wind.
- The tree's suckering is notably aggressive.

The wood is very hard and heavy, and tends to come in strange contorted shapes. It has traditionally been popular for walking sticks and for the Irish club known as the shillelagh.

Although inedible raw, sloes (as the fruit is called) can be made into a pleasantly tart jam. They are the key ingredient in the British liqueur called sloe gin. Liqueurs with the fruit are also made in Spain and Italy. Historically, sloes were used to adulterate port.

Up to 10 ft. (3 m)

Up to 15 ft. (4.5 m)

The leaves have been used for making a green dye, and the fruit a pale blue one for dyeing linen. A yellow dye can be prepared from the bark.

Douglas Fir

A large evergreen conifer that is very common along the west coast of North America, from Canada to northern Mexico. It is among the tallest of all trees, potentially to 400 feet (120 m), although none of this height are currently extant, because hugely destructive logging took place during the late nineteenth and early twentieth centuries. It has been widely introduced to other regions with a similar climate, where it has become a very successful timber tree; in New Zealand and the United Kingdom, it has even naturalized, although not to a problematic degree.

The tree has rugged bark, foliage that looks like many other conifers, but a highly distinctive cone, with a three-forked tongue poking out from below each scale. It plays an important part in the ecology of the Pacific Northwest region, with old-growth forests supporting species, such as the spotted owl, which can only survive there. Douglas fir does not tolerate shade, and so its stands do not regenerate well—the advantage it has over other dominant tree species is that it withstands fire better.

The name commemorates one of the most indefatigable of nineteenth-century plant hunters, the Scot, David Douglas, who was one of the first Europeans to see the tree and realize its potential.

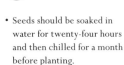

- Seeds should be soaked in water for twenty-four hours and then chilled for a month before planting.
- Germination will occur over several weeks at around 68°F (20°C).
- The species flourishes in high rainfall climates on any reasonable soil.
- Douglas fir must have full sun wherever you plant it. It has very little tolerance of shade or drought; performance on lower fertility soils will tend to be poor.

The bark of the young tree contains numerous resin blisters. If boiled down, the resin yields a tar that Native American peoples and early settlers used for caulking containers, such as jugs, and the sides of small boats in order to waterproof them.

This is the most important tree for the North American timber industry. Not only is it produced in huge quantities from sustainably managed forests, but it is strong, dense, easy to work, durable, and attractive, with reddish-brown heartwood and a distinct grain. It is used in construction, railroad ties, flooring, and for manufacturing plywood and pulp.

Up to 70 ft. (20 m)

Up to 400 ft. (120 m)

The Native peoples of the Pacific Northwest used resin collected from the trunk in poultices for treating wounds and sprains, as a chewing gum for coughs, and to soothe sore throats. It has been shown to be antiseptic.

A forest of these trees is a spectacle too much for one man to see.

DAVID DOUGLAS, SCOTTISH BOTANIST, WHO INTRODUCED THE DOUGLAS FIR INTO CULTIVATION IN BRITAIN IN 1827

> *Down the entire fruit, from the rind to the seeds. It's all edible—and nutritious.*
>
> J. BOWDEN, "THE 10 BEST FOODS YOU AREN'T EATING," MEN'S HEALTH WEBSITE

Guava

A small evergreen shrub or tree—growing to only 13 feet (4m)—the guava is a widely cultivated fruit tree, originally from Mexico and southward into South America. It naturalizes easily and has become an invasive alien in many places, most notoriously in the Galapagos Islands. The leaves have distinct veins giving them the appearance of being pleated. The fruit varies in size (from that of an apricot to a grapefruit) and shape, but tends toward the pear-like. The color of the flesh may be yellow or pink, the latter often being regarded as more nutritious. The skin is so soft that it is easily eaten, but does little to protect the fruit, which limits its commercial possibilities. It is also full of small seeds, which can be crushed when eaten, but that tend to get stuck in the teeth. There is a long history of the medicinal use of the bark and leaves.

With a constant expansion in global trade and a growing awareness of the health benefits of guava fruit (being rich in vitamin C and antioxidants), the amount of land being used to cultivate it is growing, although in many places much of the fruit sold is simply collected from the large number of trees that have gone wild. New varieties are being bred in the United States, Australia, Brazil, and India, with a high-yielding seedless one a major goal.

- Germination of fresh pits is easy at around 68–77°F (20–25°C).
- Growth is surprisingly slow for the first few months but then speeds up; flowers and fruit are possible on trees as young as two, but this is likely only in the tropics.
- The foliage is attractive enough for it to be a potential house plant for light conditions; it can also be grown as a bonsai.
- It flourishes in any reasonable soil, but is not drought or frost tolerant.

Guavas are grown as a fruit throughout the tropics and subtropics, the delicious fragrance of their ripe fruit announcing their presence. They are usually eaten fresh, although guava cheese, made by boiling the fruit with sugar, is a popular sweet preserve for the seasons when the fresh fruit is not available. The fresh fruit can also be seeded and sliced in a dessert or salad.

Guava fruit are especially rich in vitamin C, with the skin containing a particularly high proportion. The fruit is also high in tannins, phenols, flavonoids, carotenoids, essential oils, and fatty acids.

Fresh fruit, or the thick juice, is used as a treatment for gastric problems. A tea made from the leaves is used as an antiseptic wash for wounds and acne, and as a mouthwash. Research indicates that leaf products may help in treating complications of diabetes.

Based on traditional uses, guava extracts are used in cosmetics, particularly in moisturizers for skin. They are also used to fragrance and enrich hair products.

Up to 13 ft. (4 m)

Up to 13 ft. (4 m)

Pomegranate

These small evergreen trees grow to 25 feet (8 m) and bear stubby bright orange-red flowers and a very distinctive edible fruit. They have been cultivated since ancient times—with the earliest mention in Babylonian texts—and remain very popular today. So long has it been in cultivation that its natural origin is obscure, but it is probably from Iran across to northern India. The tree is attractive enough, with its somewhat twisted habit of growth and bright flowers, to be used as an ornamental; a dwarf form ("Nana") is grown as a conservatory plant in climates too cold for outdoor cultivation.

Having been in cultivation for such a long time, the pomegranate has acquired a significant body of myth and symbolism—one of the best known of which is the Greek story of Persephone being lured to the Underworld by its ruling god Hades. It has also been suggested that the fruit that Adam took from the Tree of Knowledge in the biblical Garden of Eden was not an apple, but a pomegranate. The fruit is used in cooking, baking, juices, and alcoholic beverages.

Today, pomegranate plantations are to be found across the world in suitable climate regions. India is the largest producer, but Afghanistan has the largest number of cultivars, leading some experts to suggest that the country is really the historic center of diversity.

- Pomegranate seeds should be removed from the fruit and the outer juicy coating removed, the seeds washed and then sown in compost at around 68°F (20°C).
- Germination occurs after three weeks.
- Grown as container plants in cooler climates, pomegranates should be watered as the soil dries out, and kept pruned.
- They will flourish outside in continental or Mediterranean climates, where temperatures do not go below 14°F (-10°C).

Since it is quite hard work to eat raw, much of the world pomegranate production goes into processing the seed-laden fruit into juice or various syrups. Grenadine was originally made from them, but now often contains other fruit juices. Pomegranate molasses is a usefully fruity and sour cooking ingredient much used in Middle Eastern cookery. Dried seeds are used to provide sour notes in a variety of dishes in the region.

Pomegranates feature in many religious rituals. At the Jewish New Year, Rosh Hashanah, it is customary to eat the fruit (or food containing it) and it appears on celebratory cards.

Up to 13 ft. (4 m)

Up to 25 ft. (8 m)

The bark, especially the root bark, and the peel contain considerable quantities of tannin, and this has been used for treating leather.

The fruit peel or bark is used to treat gastrointestinal problems and parasites in Ayurvedic medicine. Its astringent quality helps to stop bleeding. In Indian folk medicine, it is also used as a contraceptive. Research indicates possible health benefits to the pomegranate, particularly for the heart.

Pomegranates supposedly contain 613 seeds. Thus, Jews display their desire to fulfill God's 613 mitzvoth (commandments from the Torah) by eating the pomegranate.

GLORIA SHIMONI,
"CELEBRATING WITH
POMEGRANATES," FROM
KOSHERFOOD WEBSITE

Place names featuring the pear are manifold: Perry Barr, Perivale, Pershore, Pirton, Pyrton and Pyrfold are just a few . . .

ARCHIE MILES, *SILVA: THE TREE IN BRITAIN* (1999)

Pear

Wild pear trees are found across Europe and the Caucasus, and pears have been grown as a crop at least since Classical antiquity. Among cultivated trees there are two broad categories: those grown to produce fresh fruit, which tend to be large and juicy, and those intended for perry, an alcoholic drink similar to cider, which are smaller, harder, and more round. Trees of the former type have always been pruned and trained to maximize high-quality fruit production, whereas the latter are usually allowed to grow tall—to 50 feet (15 m). They are all long-lived trees, with perry pears achieving maximum productivity from a hundred years onward. Some are cultivated as ornamental trees.

Pear flowers are white; they are produced somewhat earlier than apples, which makes them that much more vulnerable to late frosts, and so their range tends not to go far north. The fruit is famously delicate, but modern packing and logistics ensures that producer regions are now able to export nearly ripe fruit all over the world. Serious selection and breeding started in the eighteenth century, and some old varieties are still regarded as commercially important.

- Pear seeds germinate after several weeks of cold stratification, but the resulting trees bear little resemblance to their parents.
- Commercial varieties are always propagated by grafting, usually onto a quince rootstock.
- Pears need deep and fertile soils, moist but well-drained and positioned to avoid wind or late frosts.
- Pruning restricts growth and maximizes fruit production.

Pear wood has long had a reputation as one of Europe's finest woods, and only being available in small quantities has made it all the more desirable. It is pale in color, dense, hard, non-warping, and easy to work and carve, leading to it being a wood of choice for woodcarving, and the crafting of musical and draftsmans' instruments and kitchen implements, such as spoons, scoops, and stirrers. If stained, it can be used as an ebony substitute.

Pears are popular the world over for their subtle flavor and lush texture. The flavor is not one that lends itself to preserve making, however, with one exception—the brandy known as "Williams." It is popular across central Europe and is named for a particularly juicy and productive variety first identified in eighteenth-century England as "Williams Bon Chretien" (or "Bartlett" in North America). Williams is the most common pear variety grown outside Asia.

The leaves of the pear tree were smoked in Europe before the introduction of tobacco at the end of the fifteenth century. However, the effects of doing so are not known.

Pears have a higher proportion of dietary fiber than most fruits, which is regarded as beneficial, and means that they can be used as a gentle laxative. They are also a good source of vitamin C, most of which is contained in the skin.

Up to 40 ft. (12 m)

Up to 50 ft. (15 m)

Quercus alba

White Oak

The white oak is still one of the most common hardwoods of eastern North America, and one that has been a major source of quality timber since European settlement. Growing to no more than 80 feet (25m) tall, it makes up for this by substantial sideways spread—some older trees are as wide as they are tall. The bark is gray, and occasionally almost white (hence the common name). It is a long-lived oak; some examples have been documented to be more than 450 years old. The wood has a closed cellular structure, making it water resistant, and is used for wine and whiskey barrels. White oak is thought to have declined since pioneer days; it is out-competed by other trees, partly because of the suppression of forest fires, of which it is relatively tolerant. It has simply stopped regenerating, which will have long-term consequences for North American woodland diversity.

- Acorns lose viability quickly, but germinate readily in spring if sown in fall.
- The species is notoriously difficult to transplant, so should be planted out quickly.
- Deep, fertile soils are needed, preferably acidic.
- Shows tolerance of shade when small, and of drought once established.

Hard and heavy, white oak was one of the most widely used hardwoods in history, resulting in considerable deforestation. It is very durable, and can be worked easily. Today, several closely related species are sold under its name.

The acorns have less tannin than most, and so were the easiest for Native Americans to prepare for eating. Often they would have been stored over winter in wet ground and eaten in spring as they began to germinate. Roasting or grinding into flour were among the preparation methods.

Up to 80 ft. (25 m)

Up to 80 ft. (25 m)

White oak, by U.S. federal law, has to be used for bourbon whiskey barrels. Many white wines, notably chardonnays, are also aged in barrels of white oak.

172 | WHITE OAK

Quercus ilex

Holm Oak

This long-lived evergreen oak can grow to 80 feet (25 m) and occurs naturally in a zone from the Atlas Mountains in Algeria going clockwise around the Mediterranean as far as Greece. It has been widely introduced elsewhere, but can potentially be an invasive species. The leaves are 1½-3 inches (4-8 cm) long, and narrowly lanceolate, not looking at first sight like an oak; they live for up to two years. *Quercus ilex* is the national tree of Malta and some of the oldest specimens are still found there. The tree is tolerant of a wide range of conditions, and so can be a co-dominant in forest regions, as with the Atlantic cedar in the Atlas Mountains. It also tends to exert considerable control over understory species through shading and a leaf-fall of slow-to-decay foliage.

The tree is one of the best for growing truffles. These underground and highly valued fungi form a symbiotic relationship with the tree's roots. The tree is therefore used in the establishment of truffle orchards. Recent research has made commercial production of "infected" oaks much more readily available.

The hard, heavy wood has always been used around the Mediterranean for construction, as well as for barrel making. Wagon wheels and farm equipment have been produced from it since Roman times, because its tannin content makes it particularly resistant to decay. It is also used as firewood.

Up to 35 ft. (10 m)

Up to 80 ft. (25 m)

The acorns of the tree are sweet and edible, but really only as famine food. Traditionally, they have helped to fatten pigs, which run free in holm oak forests in Spain and produce high-quality ham, such as *jamon ibérico*.

- Acorns need immediate sowing, and young trees should be transplanted into final positions quickly as there is poor tolerance of transplanting.
- Full sun is vital, and good drainage, but the species is tolerant of drought, thin soils, and wind (including, surprisingly, cold winds).
- It can be used as a windbreak and shade tree.

Oak forests provide a habitat rich in biodiversity; they support more life forms than any other native trees. They host over 280 species of insect.

"THE ENGLISH OAK," ON THE WOODLAND TRUST
(UK) WEBSITE

English Oak

One of the most widely distributed oaks, very common in its native Europe, but also introduced to central Asia and North America. This is a large-growing and long-lived species and one with a particular symbolic importance in England and Germany. Once much of northern Europe was covered in oak forest, but virgin forest is now very rare—millennia of human activity have seen extensive clearing, or repeated felling and regrowth. Old oaks can be very tall, the tallest being the Belarus "tsar oak" at 151 feet (46 m). Germany has many elderly oak trees, but the oldest in Europe are thought to be the Stelmužė oak in Lithuania and the Granit oak in Bulgaria, both possibly more than 1,600 years old. The tree is planted for its durable heartwood.

The oak is home to a vast assemblage of insects (which live on the leaves, buds, and in the acorns) and plays a major role in the biodiversity of its native territory. Typically it is a dominant in mature woodland. Botanists split the "species" in two: the pedunculate oak—*Quercus robur*—is a tree of alkaline and fertile soils, while the sessile oak—*Quercus petraea*—has an almost identical geographical range, but dominates on acidic and less fertile ground. The two often hybridize in the wild, resulting in *Quercus* x *rosacea*.

- Acorns ripen in the first year and should be sown as soon as fresh, and they will germinate rapidly in spring.
- Young trees should be planted out into prepared ground; because they are relatively slow-growing, they should be given some sort of protection against predation and competition for the first few years.
- They flourish in deep, fertile well-drained soils, but, if necessary, will survive in a wide variety of poor and shallow ones.

Historically, management of oak woodlands has often involved pollarding, a pruning system in which the upper branches are removed. This was primarily to provide a supply of fresh growth with which to feed cattle in spring, when grass was in short supply.

Oak is probably the most highly prized European timber, being hard, heavy and strong. Old oak is as hard as iron, so in the past much construction was undertaken using "green oak," which is surprisingly soft; as it dries out, it shrinks, pulling an expertly built structure together, but also creating the characterful twisting often seen in historic structures.

Oak trees were crucially important for shipbuilding in Europe until the advent of iron vessels. The rather odd shapes that mature oaks develop were invaluable as the "knees" and "elbows," which were vital in the immensely complex task of building a wooden ship; such naturally shaped pieces were much stronger than timber that had been artificially joined together. An eighteenth-century warship would have needed around 3,700 trees for its construction.

Up to 100 ft. (30 m)

Up to 151 ft. (46 m)

Cork Oak

The cork oak is a relatively small evergreen tree that has played an unusual, but major, role in the economies and landscapes of the Mediterranean region, where it is native to the western end. Growing to no more than 65 feet(20m), it is commonly grown in relatively open plantations, with an understory of grass and wildflowers, often used for grazing livestock. The main reason for growing the tree is the bark, which has evolved to be very thick, with a spongy texture. It contains a chemical called suberin which is laid down by the cells of the bark, and which is highly hydrophobic, helping prevent water loss and acting as insulation against heat, a useful adaptation in a fire-prone region. After a fire, fresh growth can emerge from beneath the bark, allowing the tree to gain a competitive advantage over other plants that have to re-establish from the ground upward.

Once the bark gets to a certain thickness, a thick layer can be cut off without damaging the tree, and the process can be repeated some nine to twelve years later, and indeed several times for over a century, making cork farming one of the most sustainable of agricultural practices. Cork oak plantations can form a rich ecosystem, but also one which is economically productive.

- Acorns should be planted fresh, with germination occurring in spring.
- Young trees should be planted out in full sun in well-drained soil. Established ones are hardy to around 1.4°F (-17°C).
- There some very elderly examples in gardens in the United Kingdom, which illustrate a good level of hardiness.
- The thick insulating bark helps protect the tree's branches during fire so that they can resprout quickly.

Cork's insulating properties are made use of in a variety of ways in construction. So superior is cork that it is even used in hi-tech applications, for example in motor and spacecraft manufacture. The tree's thick corky bark has also proved very effective as soundproofing, with cork tiles and flooring a common and flexible way of achieving this.

Cylinders of cork have been sealing wine bottles for centuries. Its cellular structure means that it is easily compressed on insertion into a bottle. Plastic and metal tops have begun to replace corks because of the risk of cork taint. However, producers have since developed methods that prevent the contamination of natural wine corks, which might lead to them making a comeback.

Up to 70 ft. (20 m)

Up to 65 ft. (20 m)

Fishermen have probably been making use of cork's buoyancy for fishing floats ever since humanity reached the shores of the Mediterranean. It is still widely used today.

Cork finds a particular use in the production of woodwind instruments, during which thin strips of the material are used to seal the gaps between components. It is also often used to make conducting baton handles.

Whoever cares for their grandchildren, plants a cork oak.

OLD SAYING IN THE ALENTEJO
REGION OF PORTUGAL

Raffia is an entirely ecological product. In effect, its extraction allows the maintenance and the regeneration of the Raphia *forests and provides revenue to local populations.*

"PRÉSENTATION," ON SOCIÉTÉ KALFANE FILS WEBSITE
(TRANSLATED FROM THE FRENCH)

Raphia spp.

Raffia Palm

Few plants exist where the length of the leaf can be greater than the total length or height of the plant itself. One of these exceptions is the raffia palm, which can grow to 52 feet (16 m) in height, but whose pinnate leaves can be as much as 80 feet (25 m) long and 10 feet (3 m) wide; this is particularly true of *Raphia regalis*, a West African species, whose leaves are the longest in the plant kingdom. The raffia palm is native to tropical regions of Africa, although just one of the twenty species, *Raphia taedigera*, occurs in Central and South America.

The inflorescence of a raffia palm—perhaps 15 feet (5 m) long—hangs down from the growing core atop the trunk, producing seed that is either dispersed by the wind or eaten by birds. The tree dies after flowering, although in a number of species this is followed by regeneration from the roots.

Raffia is an incredibly strong fiber, derived from tissue on the underside of the enormous leaves, that splits easily down the length of each strip. Northern Madagascar produces up to 90 percent of the world's traded raffia. Elsewhere, it is collected from wild trees but only traded locally. Raffia has many uses in textiles and construction.

- Germination requires a temperature of 77°F (25°C) or above, and does not occur before a minimum of six months.
- Growth is rapid in warm conditions if there is plentiful water.
- Continual high temperatures are necessary for successful cultivation.
- The species has no tolerance of cold, but established plants are usually able to survive short spells of drought.

The oil derived from the fruit is edible and used in cooking and soap making; the nutlike seed may also be eaten. A separate oil may be made from the leaves, but this is inedible and used only in the manufacture of polish.

Fibers from the raffia palm tend to be used on a local basis for making ropes, textiles, baskets, cushions and pillows, hats, and mats. The individual strands are generally not more than around 3 feet (1 m) long, but if knotted together they make a long twine that has enormous strength, with knots that are hardly visible. Sheets of raffia textile have been used as currency in some parts of Africa, a reflection of its high status. Raffia is often dyed in bright colors for contemporary craft use, or in browns for traditional products.

Traditionally, raffia was the gardener's string of choice for tying up plants. Although this use has gone into decline, gardeners still very much favor it for tying together grafts: it has just the right combination of strength and "give" for the purpose, it is immensely pliable, and it does not dig into soft plant tissue.

Raffia leaf stalks can be used in lieu of wooden poles for roof construction. Leaves are used for roofing, the leaflets tied to laths or leaf stalks and functioning as tiles.

Up to 44 ft. (13.5 m)

Up to 62 ft. (16 m)

Salix alba

White Willow

The white willow is a deciduous tree that can grow to 100 feet (30 m) and is very common in its regions of origin—western Europe across Eurasia to the borders of China. Full-sized trees tend to be rare, partly because of the general willow habit of falling over in wet soils and re-rooting themselves, but also because many are pollarded, to provide wands, the raw material for basket making and weaving. Willow stems less than a year old are so supple that they can be woven into wicker, a material that combines strength with an extraordinary flexibility and lightness. Pollarded willows will live for longer than unpruned ones, and can be very distinctive features of the low wetland landscapes where they usually grow, their trunks very stout and with a swollen head from which emerges a mass of straight stems.

In the landscape, these are easily recognized trees, as they have silvery-white leaves, making them obvious from afar. There is enormous variation in the species, with the color of the young bark being one of the most immediately obvious features—everything from yellow to red can be found. Trees that have been kept pruned display these colors to bold effect in winter, and can be used ornamentally.

- Like most willows, these are almost ridiculously easy to propagate—stems plunged into the ground will root over the winter, with vigorous growth from spring onward.
- Full sunshine and fertile soils are necessary for successful growth, and while they flourish on moist soils, they will grow well on any soil that does not suffer regular drought.
- Pruning is usually done in late winter.

Those trees with no side-shoots are the best for basket making; those with shoots can be used for larger and heavier structures. As well as baskets, traditional cultures across Eurasia have used wands for making fishing traps, livestock pens, primitive boats (such as the round Welsh coracle), fences, and components for houses. The fibers in the wands have even been extracted for making rope or paper.

Sap from the bark of willow contains a chemical —salicylic acid—which is effective as a pain reliever, a quality that has been recognized in almost every place where willows grow. Greek and Roman physicians knew that the bark could ease pain and reduce fever, as did the ancient Egyptians. Nineteenth-century chemists synthesized a version (acetylsalicylic acid), which is commonly known as aspirin.

Up to 70 ft. (20 m)

Up to 100 ft. (30 m)

The ability of willows to root and grow quickly has often been used by traditional cultures to help secure unstable river banks and prevent erosion. This is now being revived as an alternative to using concrete embankments.

Willow makes very good charcoal, and has been particularly appreciated by artists as a drawing material. The best quality artists' charcoal is still made from white willow twigs.

I've attacked that old giant of a pollard willow and I believe it has turned out the best of the watercolors.

Salix viminalis

Osier

A deciduous tree or large shrub, the osier continually regenerates itself from the base, a useful adaptation for the unstable habitats such as riverbanks, where it is naturally found. Given how useful a tree it is, the species has been spread well beyond its original homeland, which is now unclear, but it is thought to cover much of continental Europe and western Asia. It does not vary much, but it crosses easily with other willows—hybrids may be grown for very particular craft uses.

Superficially similar to white willow, the leaves are less silvery and longer, and the bark is gray. It has a tendency to produce very long straight stems from the base, known as withies or rods, coarser, thicker, and with less flexibility than the wands of white willow (see p.180). Traditionally, these are produced by coppicing the trees rather than pollarding them.

Osier wands tended to be used for jobs around the home or farm rather than making quality products for sale. In northern Europe, large rough baskets for home use would often be made from them and larger functional objects, such as hurdles—a woven modular fencing used for creating temporary animal pens around the farm.

- Just as with other willows, buried stems will grow rapidly into new plants.
- The trees tend to become ungainly with age and are best cut down to the base every few years, even if the wands are not needed.
- Flowers appear as catkins in early spring; male (yellow, oval-shaped) and female (longer, more cylindrical) catkins on separate plants.

Osiers have often been favored for providing the upright elements in basket making, with narrower and more supple wands from other species woven between them.

Osier bark was used for tanning in Russia until industrialization, which possibly accounts for the historic high reputation of Russian leather. Osier bark grown in Russia has been found to contain up to 8–13 percent tannin.

Up to 25 ft. (8 m)

Up to 35 ft. (10 m)

Toothbrush Tree

Found throughout the Middle East and Africa, this small deciduous tree or large shrub grows to no more than 23 feet (7 m) and has a mass of rangy branches covered by small, slightly fleshy leaves. The whole plant has a pungent and quite pleasant odor. It is found along river banks in desert areas where ground water is available, and can flourish even when the water is salty. Historically, the tree was spread by Muslim merchants along trade routes; in some Muslim countries it is common as a cemetery tree.

Relatively inconspicuous flowers are followed by red fruits that are around ½ inch (1 cm) across, the flesh surrounding a single seed. The fruits are edible, have a sweet flavor, and can be made into a drink. The leaves are also edible, with a peppery flavor, and can be fed as forage to livestock, although it is said that milk from animals fed on them can be tainted. The tree's ability to grow in salt-contaminated soils makes it a useful species in agroforestry.

- Fresh seed needs to be soaked in water for twenty-four hours before sowing, when germination at 68–77°F (20–25°C) should be fairly rapid.
- Established plants can be expected to survive light overnight frosts and seasonal drought, and be tolerant of high temperatures.
  - High humidity and constant damp conditions can be damaging.

Known as *miswak*, the twigs have had a long history of use as chewing sticks to clean the teeth and freshen the breath. Research indicates that chemicals in the wood have an antibacterial effect. The prophet Muhammad is recorded as having promoted its use.

Salvadora persica has various medicinal uses. Preparations made from the roots, the bark, and the seed are used for making salves for treating a variety of conditions such as rheumatism and other painful conditions. Pilu oil is used similarly.

Up to 25 ft. (8 m)

Up to 23 ft. (7 m)

An oil, known as pilu oil, can be extracted from the seeds of the tree. In India this provides the basis for soap making in villages in desert areas.

> *Hack a large [elder] tree back to the ground and, like Medusa, it springs anew with even more stems and added vigor.*
>
> ARCHIE MILES, *SILVA, THE TREE IN BRITAIN* (1999)

Elder

Hardly qualifying as a tree, elder is usually a large shrub, to 20 feet (6 m), and occasionally taller. It continually regenerates from the base, but if these lower shoots are cut away or continually nibbled by livestock, a single trunk can develop, and the resulting tree develops an appearance of considerable antiquity.

The leaves are pinnate, with five to seven leaflets and an unpleasant smell, heads of white flowers appear in early summer and are followed by soft black berries in the autumn. The very weak twigs and branches are filled with lightweight pith, an adaptation that allows height of stem to be traded off against weight, for this is a rapidly growing pioneer species, whose seed is spread by the birds, which drop the berries into gaps in existing woodland, along woodland edges or into waste ground. It tends to be short-lived.

This is a very widely distributed species, found across Europe, North Africa, and into western Asia. It has three subspecies in North America, one restricted to the west, another to Mexico, and one to the eastern seaboard. Unusually for a temperate zone tree, it has naturalized in parts of tropical Africa.

- Seed is best sown in fall, or given stratification treatment.
- The plant is also very easy to propagate from cuttings.
- Young plants grow rapidly in any fertile soil in sun or light shade.
- A somewhat untidy grower, coppicing helps keep it more manageable.
- A number of dark foliage cultivars are available, usually with attractive pink flowers.
- Fruit appears as glossy dark berries in late fall.

Various preparations made from all parts of the elder have been used in Native American and Eurasian folk medicinal traditions. A tea made from the flowers is used to treat coughs and colds, while some experimental evidence suggests that elder preparations may be of genuine help in hastening recovery from influenza. Dried flowers are simmered in water for fifteen minutes to produce a flavorful tea. European herbalists have tended to use the inner bark.

The dark blue-purple berries can be eaten when fully ripe, but are mildly toxic until they are cooked. They are sometimes added to other fruit such as apples and blueberries, in pies, or used in the making of jams, jellies, and chutneys. The flavor is poor and the addition of the fruit is primarily for the intense color it imparts.

The flowers have a subtle, fresh odor, which in several European cuisines can be preserved, usually as elderflower champagne or as a cordial. The flowers can be dipped in batter and fried.

Elderberries are sometimes made into a wine in Europe, on a craft, rather than commercial basis. The Italian liqueur sambuca includes elder flowers among its ingredients.

Up to 20 ft. (6 m)

Up to 20 ft. (6 m)

Sandalwood

The aroma of sandalwood cannot be mistaken for anything else. Cultivated and traded since ancient times, this has become one of the most sought-after and highly prized aromatic woods. Historically, various cultures have placed great significance on its aromatic and medicinal properties. A small and relatively spindly evergreen tree, it is a semi-parasite, its roots latching onto those of other trees (it is not particularly discriminating about which ones) and extracting nutrients. This tropical species is native to parts of India, but closely related examples are found throughout Southeast Asia and northern Australia; all can be commercially exploited for timber and essential oil, although the quantity of oil that can be extracted varies.

True Indian sandalwood is very much an endangered species. Plantations of it have even been raided by bandits. Larger trees are regarded as government property in some Indian states and there is strict control over its exploitation with a ban on its export. The species is also threatened by degradation of its habitat. Commercial production there and in Australia is increasing the supply, as, unlike many parasitic plants, it is not difficult to grow and is long-lived (up to a hundred years).

- Sandalwood seeds are large and need to be split before they will germinate; this can be done by soaking in 1 percent household bleach for twenty-four hours.
- Seedlings grow rapidly and need to make contact with a suitable host within a year to survive.
- Commercial practice is to sow the seed among existing suitable hosts.
- Harvesting is not regarded as economic until the trees are at least seven years old.

Sandalwood products have been widely used in folk medicine as well as in the ayurvedic tradition, for a wide variety of ailments. Research does indicate antibacterial properties, however, no particularly important potential uses have emerged.

Sandalwood trees are not large, but the timber is dense and of high quality, quite apart from its rich, perfumed scent. The wood has traditionally been used for high-value items, such as religious images for temples, chess pieces, and small boxes.

Sandalwood has been an essential ingredient in many Eastern incenses and fragrances for millennia. It is equally important in today's cosmetic industry. Since there are limited supplies from this species, oil from other sandalwoods and other genera is often used.

Up to 30 ft. (9 m)

Up to 35 ft. (10 m)

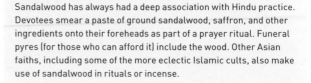

Sandalwood has always had a deep association with Hindu practice. Devotees smear a paste of ground sandalwood, saffron, and other ingredients onto their foreheads as part of a prayer ritual. Funeral pyres (for those who can afford it) include the wood. Other Asian faiths, including some of the more eclectic Islamic cults, also make use of sandalwood in rituals or incense.

> *Vishukumar, who knew that agriculture in India is a gamble with monsoons, decided to make up his loss by growing sandalwood. . . . Now, he has 1,800 sandalwood trees.*
>
> "TARIKERE FARMER HAS 1,800 SANDALWOOD TREES TO HIS CREDIT," *DECCAN HERALD*, AUGUST 15, 2014

Sassafras is considered a
good choice for restoring
depleted soils in old fields.

M. M. GRIGGS, "SASSAFRAS,"
ON U.S. FOREST SERVICE WEBSITE

Sassafras

Sassafras trees can be immediately recognized by their leaves, which vary considerably in shape, being either one-, two-, or three-lobed. Growing to around 65 feet (20 m), this medium-sized deciduous tree is common in the eastern states of the United States, from southern Maine and southern Ontario west to Iowa and south to central Florida and eastern Texas. A pioneer species, it tends to develop in gaps left by the loss of larger trees, and is then usually out-competed by larger species in time. However, it has the ability to produce suckers from the roots, and dense thickets can build up as a result. It is most often seen as a relatively low shrubby understory, growing beneath a mature woodland canopy. The fruit is a blue-black drupe containing a single seed that is dispersed by birds.

A distinct spicy fragrance can be detected on all parts of the sassafras plant; the active ingredient is safrole, which makes up most of the essential oil that can be distilled from its root bark or fruit.

Sassafras has some popularity as a garden and landscape plant, largely because of its attractive foliage, which turns yellow and orange in fall, and its aromatic scent.

- Sassafras seeds germinate well in spring following an fall sowing; alternatively a four-month stratification is needed.
- Young plants develop a taproot which, if broken, is liable to result in the death of the plant; they need to be planted out quickly.
- Transplanting is regarded as near impossible, but suckers may be used for propagation.
- Saplings are tolerant of some shade, but not older plants. Acidic soils are preferred.

Native American and early settler herbal practice made considerable use of sassafras, largely as a tonic. It has also been employed against toothache as a pain reliever and as an effective repellent for mosquitoes and other insects.

Root beer is a drink, which may be prepared to be either alcoholic or nonalcoholic. It dates back to the early days of white settlement of North America. It was prepared using extracts from the bark or the roots of sassafras, as well as other flavorings. It is now made with other plant extracts.

Gumbo filé is a powdered spice, used in Cajun cuisine as a flavoring and thickener for stews and sauces. Made from the leaves, it does not contain appreciable quantities of safrole and is therefore regarded as not posing a health risk.

Safrole has been regarded as a carcinogen and banned from food products in the United States since 1976. "Sassafras" products now replace it with methyl salicylate, which has a similar aroma. Safrole is also used to make illicit drugs, such as MDMA.

Given its distinct and pleasant fragrance, it is not surprising that sassafras extracts were used in making soap and cosmetics in the past.

Up to 60 ft. (18 m)

Up to 65 ft. (20 m)

Senegalia senegal syn. *Acacia senegal*

Gum Arabic Tree

Gum arabic is an extremely useful material for a wide range of purposes, and so far there have been few synthetic alternatives. There is an irony that this specialized, but essential, raw material is still collected by hand from wild plants by some of the poorest people on earth in some of its most difficult ecological and political environments.

Found across the sub-Saharan Sahel region of Africa and the southern part of the Arabian Peninsula, as well as in Pakistan and northern India where it has been introduced, the gum arabic tree grows to a maximum of 40 feet (12 m) and looks like many other acacias, with dense branches covered in large thorns and small evergreen leaves. If cut, the trunk oozes a resin that hardens to form gum arabic. Traditionally this was collected by nomadic farmers and then sold to middlemen who got it onto the international market; increasingly, however, it is being grown in plantations.

Gum arabic dissolves easily in water, and reduces its surface tension. As it dries, it forms a transparent coating. It works well as an emulsifier and is nontoxic, so has a particularly important role in the food industry. All this adds up to a unique set of properties that make it more or less irreplaceable for many purposes, although like all natural materials its composition and quality varies unpredictably.

- Seed of the tree needs soaking in hot (almost boiling) water for ten minutes before it is sowed.
- Germination is then relatively rapid.
- The tree may also be propagated by cuttings.
- Sandy alkaline soils are generally preferred.
- There is tolerance of heat, salinity, and drought, but not of prolonged cold or damp.
- Trees on infertile soils tend to produce more gum, which is drained from cuts in the bark.

Gum arabic is an essential ingredient in watercolor paints, because it delays drying, and then binds the pigment to the paper thus improving its luminosity. It is also used in photographic printmaking and in various printmaking processes in which it is used to protect and etch an image. It is also used as a water-soluble binder in the manufacture of fireworks.

Although not a traditional food item, gum arabic's qualities as an emulsifier and thickening agent means that it is used extensively in a comprehensive range of food products, including ice cream, confectionery, baked goods, jellos, and imitation dairy products. It is also used to increase the fizz in carbonated drinks, a factor that has led to threats by Sudan (the largest producer) to pressurize the government of the United States by limiting exports.

The nontoxic adhesive properties of gum arabic are exploited for the sticky surfaces of stamps and cigarette papers.

The surface roots combine immense strength with flexibility and are widely used for making rope, string, and nets throughout the tree's range.

Up to 53 ft. (16 m)

Up to 40 ft. (12 m)

In May 2013, more than 60 people died fighting over gum arabic land in Darfur . . . continuing civil conflict, fueled in part by the high demand for the gum itself, keeps farmers vulnerable.

MARIE VON HAFFTEN,
"THE CURSE ON GUM ARABIC
IN SUDAN," GLOBALENVISION
WEBSITE

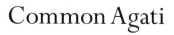

Common Agati

A small evergreen tree growing to no more than 50 feet (15 m), this species is originally from India and Southeast Asia, but it is now widely grown throughout Africa. The flowers are white, and the leaves have the pinnate form typical of the pea family. This is a tree that may be short-lived (it is so shallow-rooted that it is easily blown over). The wood is of poor quality, but the tree is becoming increasingly popular in agroforestry because it is particularly fast-growing and can be coppiced or pollarded. The leaves make good forage and up to eight cuts of foliage may be harvested every year, a boon for poor farmers. An environmental adaptation that gains the tree further favor in agroforestry is its ability to cope with seasonal flooding—in waterlogged conditions new roots may even be produced at a higher level up the stem.

The wood of the common agati is soft and has little practical use beyond supplying poles, which are used to support climbing plants, such as beans or pepper vines. The wood is also used for making temporary fencing.

The leaves, immature seed pods (which look like beans), and flowers are eaten as a vegetable in many regions of India, Sri Lanka, and Southeast Asia, including Thailand and Laos. They are usually steamed or included in currylike stews.

Agati is mainly grown as a forage crop, with foliage being cut and fed to cattle, especially in dry seasons. It is less palatable to other livestock. Its light growth, however, casts little shade so other crops can grow around it, making it useful for small-scale farm use. Its foliage is used as green manure and compost.

- Seeds should be soaked in warm water for twenty-four hours, after which germination should be rapid.
- This is a tropical species needing mean minimum temperatures of 71–86°F (22–30°C), and there is no tolerance of cold.
- Agati is best suited to moist climates, but can be successful in dry conditions.
- Growth is very rapid in the first year, then slows.

Up to 10 ft. (3 m)

Up to 50 ft. (15 m)

Shorea robusta

Shala

A relatively slow-growing tree, reaching 115 feet (35 m), and sometimes even taller, *Shorea robusta* makes up around 14 percent of the forest cover in its homeland of northern and eastern India. The shala (or sakhua) tree has long had an association with the Hindu god Vishnu, while other Indian religious traditions, including Buddhism and Jainism, also feature the tree in their mythologies. Its leaves make it instantly recognizable, being 10 by 16 inches (25 x 15 cm) and with a stiff texture. The tree can be evergreen, but in seasonally dry climates it is deciduous. The name is derived from Sanskrit and suggests that it has been used for building houses since ancient times.

Shala leaves are so tough when dried that they make good impromptu plates. In northern India they are used to make *patravali*—disposable and biodegradable plates or bowls made from several leaves and sown together with tiny pinlike sticks.

A heavy, coarse-grained hardwood, shala timber is ideally suited to construction, framing, and shipbuilding. It does not plane well, so therefore tends not to be used for interior or detailed work.

Resin extracted from the wood of the shala tree is used in making incense for Hindu temple ceremonies, while the oil expressed from the seed is used to light temple lamps. The tree resin is also employed as an astringent in Ayurvedic medicine.

• Seed loses viability quickly, within two weeks.
• In commercial plantations, fresh seed is sown directly into the ground.
• Warm moist climate conditions are needed although there is tolerance of light overnight frosts, and seasonal drought.
• Shala flourishes on any soil between heavy clay and sand.
• The tree has a great ability to regenerate from the stump after fire damage.

Up to 50 ft. (15 m)

Up to 115 ft. (35 m)

Rowan, Mountain Ash

A small, upright tree—its maximum height is 50 feet (15m)—the rowan is commonly found in mountain regions across Europe and North Africa. A member of the rose family, its pinnate leaves and flat clusters of white flowers make it distinctive in spring, while the orange-red fruit is often the first sign of fall. The species is highly variable, with some five regional subspecies, and botanists have long had problems defining it. As well as some cultivars developed for fruit, a number have been selected for use as garden and landscape plants, the small size and ability to cope with poor soils making them good trees for urban areas.

Rowan is a short-lived pioneer species and seedlings can often be found emerging out of rock faces and other inhospitable places. This undemanding species is able to grow at relatively high altitudes and copes well with exposure and poor soils.

- Seed germinates in spring after a fall sowing; alternatively six weeks stratification may be tried.
- Seedlings are best grown in pots for a year before planting out.
- There is little tolerance of waterlogging or shade.
- Named cultivars have to be propagated.

Although not much used now, rowan once had a high reputation in Britain, as its timber was all heartwood. It tended to be used for tools and small, relatively high-value items. It was also valued for the fact that it could also be stained almost any color.

Alcoholic spirits based on rowan berries, including wine and liqueurs, have a long history in Austria, parts of Germany, and the Czech Republic.

The berries are unpalatable, but can be made into a tart jelly that is good for accompanying game. Since the eighteenth century there have been efforts in Russia and Moravia to develop tastier varieties.

Up to 23 ft. (7 m)

Up to 65 ft. (15 m)

Sorbus torminalis

Wild Service Tree

The wild service tree (sometimes known as the checker tree) is widely distributed and found across most of Europe, northwest Africa, and the Caucasus. However, it is a species that is never abundant and consequently rarely noticed. Its patchy distribution and questions over its former uses create something of an air of mystery around it. This medium-sized deciduous tree grows to a maximum of 100 feet (30 m) and is most easily noticed in the fall when it turns yellow to red-brown and produces bunches of dark red fruit. It produces flowers in late spring to early summer that have five white petals. The leaves, uniquely for a *Sorbus*, are shaped a little like maple leaves and are dark green on both sides. Across some of its range at least, its ability to regenerate through seed is limited and it survives through its ability to throw suckers.

- Seed needs at least two months cold stratification or fall sowing.
- Suckers, cuttings, and grafting are also possible.
- Most well-drained soils are suitable for this tree. It will grow equally well in alkaline, neutral, or more acidic soils.
- It can tolerate some shade.

There appears to be a strong association between the tree and alcohol and drinking houses in late medieval England, but it is not clear today how the fruit was used. A schnapps is still made from it in Austria.

The wood is among the hardest of all European timbers, combining strength with elasticity, and it is an attractive pink color. It is very similar to pear wood, and there is no doubt that service tree wood has often been mislabeled. Being hard and strong, it was much favored for gun and crossbow stocks, parts for musical instruments, and for tool handles.

Up to 50 ft. (15 m)

Up to 100 ft. (30 m)

Until sugar was imported in the eighteenth century, the fruit appears to have been popular across Europe, most probably as a form of sweet snack.

Of all the botanical wonders discovered in the New World by the first European explorers, few can compare with the benzoin plant for its fascinating history.

Y. PURWANTO ET AL., "THE ETHNOBOTANY OF BENZOIN [STYRAX SPP.] IN SUMATRA," *JOURNAL OF TROPICAL ETHNOBIOLOGY*, 2:1, JANUARY 2005

196 | BENZOIN TREE, SUMATRA BENZOIN

Benzoin Tree, Sumatra Benzoin

This is a small tree native to Sumatra growing to 40 feet (12 m), which is the principal source of benzoin resin—an aromatic material that has been a precious commodity for many centuries. There are, in fact, around 150 species of *Styrax*, distributed throughout Eurasia and North America. Formerly, the Middle East and Europe obtained its resin from *Styrax officinalis*, a Mediterranean species, however, from the fourteenth century onward, traders started to supply the resin from other species found in the rain forests of Southeast Asia, most notably from *Styrax benzoin*. Other species of *Styrax* in Southeast Asia are also tapped, but this is regarded as producing the best quality.

The resin is made by collecting the dried fluid that oozes out of cuts that are scored in the bark of the tree; like many other tree resins, its purpose is probably a defensive one—to deter sap-sucking insects. Benzoin trees have been cultivated in plantations to supply demand, probably for hundreds of years, but they are also often integrated into a forest-garden-style of small-holder agriculture. The future is likely to see more systematic selection and cultivation of high-yielding varieties of *Styrax* in order to produce resin more sustainably. The tree is also grown ornamentally for shade in Africa.

- The seed needs to be sown fresh, while it is still moist.
- Young plants are transferred to nursery beds before planting out in their final positions.
- Much cultivation is now happening in areas where soils have been depleted by the destructive practice of shifting agriculture, so fertilizer application is often necessary for good growth.
- Tapping for resin can be done from six years onward, through incisions made at the base of the tree.

Styrax wood is used in the countries where it is grown for light construction. A related species, *Styrax japonicus*, supplies wood for the manufacture of traditional musical instruments in Japan.

Balsamic benzoin resin obtained from the bark of several *Styrax* species has been highly sought-after for making incense and perfume for centuries, if not millennia. Not only does it have an agreeable fragrance of its own—a sweet, vanilla-like aroma—but it also acts as a fixative, bonding to other scents and slowing down their dispersal into the atmosphere, which makes it extremely useful. It is similar to, and has often been confused with, frankincense, another plant-derived resin.

Although little research has been done, claims for the resin's medicinal value have a long history. Islamic cultures used it for various ailments, while it is an ingredient in Friar's Balsam, a traditional medicine used as an antiseptic for cuts and for inhaling to alleviate cold symptoms.

Benzoin resin has been used in the past (together with rose water) as an ingredient in "virgin's milk," a prized French beauty treatment.

Up to 8 ft. (2.5 m)

Up to 40 ft. (12 m)

Mahogany

Mahogany is an evergreen tree that grows to a maximum height of 80 feet (25 m). Several related species produce highly esteemed timber, but *Swietenia mahagoni* is the one that first established the wood's reputation. In the sixteenth century, Spanish invaders of the Caribbean realized the value of the tree after seeing the native people of the region carve their dugout canoes from its trunks. With its high status and slow growth, mahogany was intensively exploited, and by the early 1920s it was one of the first timbers to be seriously depleted; other *Swietenia* species were then exploited. Illegal importation of mahogany into North America was widespread until the early 2000s. Mahogany can be grown in plantations, but its slow growth has discouraged efforts to do this on a large scale.

- Mahogany seed germinates at temperatures of 68–77°F (20–25°C), taking up to eight weeks to do so.
- Commercially, seedlings are transplanted to nursery beds, where they are put out to their final planting sites.
- It is possible to grow mahogany as a bonsai.

Researchers in Egypt have looked into using a tincture of leaf and bark extracts against the varroa mite, a serious parasite of honeybees.

Mahogany was long esteemed as one of the world's finest timbers, particularly for shipbuilding during the age of sail, because it would resist decay and absorb cannon shot and other damage without splintering. It was favored by European imperial powers for their battleships, and is still used for building naval minesweepers, which need to be non-magnetic.

Much fine nineteenth-century furniture was made from mahogany, some of it recycled from obsolete wooden warships. Today, its fine and even grain and good sound resonance make it popular for making speaker cabinets, in addition to high-quality scientific instruments.

Up to 46 ft. (14 m)

Up to 80 ft. (25 m)

Tabebuia impetiginosa

Pink Ipé

Highly noticeable when in flower, this deciduous species grows to 100 feet (30 m), and is most familiar as a street tree in countries with warm climates, although originally it was from the tropical regions of South America. A slow-growing tree, it produces very dense timber that is strong and saturated with tannins, qualities that make it resistant to rot and therefore very desirable commercially. The species' profusion of bright pink tubular flowers, each 8 inches (20 cm) long, has led to its being widely planted as a street tree. The flowers are often borne at the end of the dry season when the branches are leafless and bare.

Pink ipé has been one of the most sought-after of tropical hardwoods and its exploitation has been the cause of some of the worst destruction of lowland rain forest in South America.

Indigenous people in South America use the inner bark to treat a number of ailments. It is an effective antibacterial agent, and Western herbalists have prescribed it as a general tonic. However, it contains potentially harmful toxins, and scientists warn that it should not be used internally.

Up to 115 ft. (35 m)

Up to 100 ft. (30 m)

Pink ipé is so hard, heavy, and difficult to work that its uses are limited to construction, particularly for applications where there will be exposure to rain, such as decking. The timber is a very good load-bearer. All supplies are currently from the wild, and so in reality are unsustainably harvested.

Looking for ipe is like following a trail of stumps.

MARCO LENTINI, TECHNICAL MANAGER, INSTITUTO FLORESTA TROPICAL, BRAZIL, QUOTED BY L. SPECKHARDT IN "A TRAIL OF STUMPS," ON *LANDSCAPE ARCHITECTURE MAGAZINE* WEBSITE

- Germination occurs at 86°F (30°C).
- The tree has a relatively small root system, making it a good choice for urban environments in warm-climate zones, and also containers or sites with confined root runs.
- Young trees grow quite quickly, slowing as they age.

In Malaya there is the belief that tamarind makes one wise, so a little tamarind with coconut milk is placed in the mouth of an infant at birth.

JULIA F. MORTON, "TAMARIND,"
IN *FRUITS OF WARM CLIMATES* (1987)

Tamarind

Tamarind is a leguminous evergreen tree that can grow to around 100 feet (30 m) and develop a considerable spread. It was originally from tropical Africa but was introduced to Asia, most probably more than 2,000 years ago. The fruits of the tamarind were widely traded in ancient times. In Asia, where it has been most widely used and extensively cultivated, the word "tamarind" is derived from the Arabic for "date of India." European colonists introduced it to Mexico and then on to the rest of South America in the sixteenth century. Today, tamarind is an important source of flavorings for cuisines in warm-climate countries, in particular in the Indian subcontinent and Mexico. Tamarind is also used in traditional medicine.

The leaves are divided into many leaflets, and, as is often the case with woody pea family species, close up at night; they are highly acidic, enough to rot fabric left in contact with them. Insignificant flowers mature to stout pods containing hard seeds immersed in soft, juicy pulp; it is this pulp that is the source of the food ingredient tamarind. Relatively little vegetation will grow beneath the tree, and this may have led to the widespread belief that it is harmful to sleep beneath one.

- The seed is very hard and should be soaked in water that has just been boiled to soften the coat.
- Germination, at around 77°F (25°C), should then be rapid.
- Plants need warm conditions, but established ones will survive very light and very brief frosts.
- The tree is tolerant of a wide variety of soils, and will stand salt spray well.
- Tamarind makes an excellent bonsai specimen.

Tamarind pulp has a pleasantly fruity sour flavor, one that lends itself to an amazingly wide range of uses. Its most basic use is simply as a flavoring, but it is also used to make chutneys and pickles; despite the sourness, it is also used for jams, ice creams, and desserts. The flavor is refreshing and a drink based on mixing the pulp into water is popular in India, Mexico, and many other countries with hot climates. The flavor varies between varieties—some are sweet and can be eaten fresh or used in making candies.

The pulp, leaves, flowers, and bark are used in many traditional medical systems. In particular, they are used to bring down fevers and treat digestive disorders. Tamarind pulp has also been used in a poultice against skin inflammation.

The acidity of the pulp makes it useful for removing tarnish and the greenish patina on copper and bronze objects. In Buddhist Asian countries the pulp is used to polish brass shrine statues and lamps.

Dense, durable, termite-proof, and with an attractive red coloration, the timber is used for furniture and any objects that receive considerable wear and tear, such as rice pounders and mortars. It is also good for flooring and furniture.

Up to 105 ft. (32 m)

Up to 100 ft. (30 m)

Taxodium distichum

Swamp Cypress, Bald Cypress

One of the most distinctive conifers of North America, the swamp or bald cypress is a deciduous species found in wet habitats along the southern and southeastern seaboard of the United States. The swamps form a unique ecosystem and are renowned for their rich animal diversity. The tree's exceptional ability to survive long periods of inundation results in it dominating large areas, although over much of its habitat in the Mississippi Delta, the incursion of seawater is killing extensive tracts. Growing to 130 feet (40 m) and with its feathery pale green foliage, this is an attractive tree, which has been extensively planted as an ornamental well beyond its homeland.

An intriguing aspect of the swamp cypress are its "knees," technically pneumatophores, which are woody structures that emerge from the trees' roots, often some distance from the trunk. It is generally believed that these also allow oxygen to get to the roots, although expert opinion about their function is still divided.

Felling on a huge scale in the nineteenth century has resulted in a major loss of swamp cypress forest, and the destruction of some very large and old trees. The oldest alive today is in North Carolina, and is thought to be more than 1,600 years old.

- Seeds need fall sowing or stratification for two months.
- The seed bed and compost for the first few months needs to be moist, but not waterlogged.
- Seedlings are generally grown for a year in a nursery bed or container before planting out in a permanent position.
- Young trees need sunlight and a fertile moist soil.
- Growth can be slow, particularly in cool summer climates. It is a good bonsai.

Swamp cypress timber is highly regarded, particularly that of older trees, which are saturated in compounds that resist fungal decay very effectively. The trees will grow well at high density, making cypress forest one of the world's most productive ecosystems, and a good prospect for commercial forestry. One of the most common uses for the timber is in the production of roofing shingles.

Native Americans used the trunks to make dugout canoes. Examples, up to 4,500 years old, have been uncovered during earth-moving operations in North Carolina. Today, cypress planks are used for making small craft and the wood taken from some swamp areas is prized for special uses, such as wood carving.

Up to 100 ft. (30 m)

Up to 130 ft. (40 m)

"Pecky cypress" is wood from older trees that has been infected by a fungus, which causes small holes to develop, and a rich coloration. This unique wood is popularly used for interior paneling, and for other decorative features.

The sticky resin from the cones of the swamp cypress tree was used by Native Americans as a treatment for wounds and skin abrasions. The resin blends well with many aromatherapy oils.

> There is no finer scene than
> that of a huge old-growth bald
> cypress tree standing proud in a
> blackwater swamp . . . the stately
> bald cypress epitomizes the Old South.

KEVIN ADAMS, "BALD CYPRESS TREES *TAXODIUM DISTICHUM*
IN NORTH CAROLINA" ON KADAMSPHOTO WEBSITE

Written evidence for yews as sacred trees comes from the Christian laws of Wales, which require compensation equivalent to sixty sheep for destroying a "saint's yew."

OLIVER RACKHAM, *THE HISTORY OF THE COUNTRYSIDE* (1986)

Yew

Yew is a conifer, but one that is very different to other conifers. Its wood is very much harder than most, and it does not have cones, but red juicy structures that resemble berries. It also grows in a way that is different to most trees—a cycle of growth and regeneration that takes centuries to complete, and that means that it can potentially live for ever. There are some very elderly trees in Europe, but because the central trunks have rotted away it is very difficult to work out how old they are. Some are almost certainly at least 2,000 years old; other estimates put a few specimens at twice that age.

Growing as high as 80 feet (25 m) and generally equally wide, yews are densely clothed in soft, very dark green needles, with increasingly wide trunks as they age. They have an immense capacity to regenerate from cuttings, which has made them highly popular as hedging, and as subjects for topiary.

There are nine species, scattered around the Northern Hemisphere; botanists disagree about the extent to which they should be regarded as separate. *Taxus baccata* is found across much of Europe, North Africa, and western Asia.

- Seed needs to be sown in fall and left out for two winters before germination can be expected.
- Much propagation is done from cuttings, however.
- The trees grow moderately quickly when young.
- They are tolerant of some shade, on dry and alkaline soils, but are rapidly killed by waterlogging or excessive moisture.
- They are very easy to keep trimmed into almost any desired shape.

Yew wood is not produced in any quantity or with any great speed, so it only tends to be used for special applications. It is among the hardest of the softwoods and is easy to work. It is very beautiful, with a deep luster, pinkish coloration, and a large number of knots, all of which make it ideal for carving, decorative surfaces, and other high-value applications.

Yew is associated in the popular imagination with Europe's various pagan traditions and mythologies. In some places, the Christian church took over the yew as a symbol of immortality and transcendence, and it is therefore commonly found in churchyards and church cemeteries across the United Kingdom, France, and northern Spain.

Yew is infamously poisonous, the major toxin being alkaloid taxine. The flesh of the berries is thought safe, but if the seed is eaten death may result. Horses and cattle are not infrequently poisoned by eating the foliage.

Yew foliage is a source of taxol, a compound used in cancer chemotherapy; the Pacific yew (*Taxus brevifolia*) is a particularly rich source. It is now possible to produce taxol through biosynthesis instead.

Up to 65 ft. (20 m)

Up to 80 ft. (25 m)

Tectona grandis

Teak

A distant relative of garden mint, the teak tree is a deciduous species growing to 130 feet (40 m) from the tropical regions of India, Southeast Asia, and Indonesia. There is a huge global demand for teak timber, which is very strong and saturated with oils that waterproof it and help prevent the growth of decay-causing fungi. Fortunately the tree is easy to propagate and grows quickly so teak plantations have been common since colonial days.

Breeding teak plants that grow faster and straighter and yield better wood is a slow business, largely because of the long wait before the trees flower. The rate of seed set and germination can also be slow.

Teak wood is most familiar to people outside the tropics as a material commonly used for outdoor furniture, its red-brown color weathering to a stately silver-gray over the years. Being saturated with oils helps to keep water at bay and makes treatment with preservatives unnecessary.

In India and the Philippines, a tea made from the bark or the leaves is considered useful in treating fevers, headaches, and digestive difficulties.

• Seed needs several days of alternate wetting and drying before germination, which can take up to two months.
• The tree does best on deep, alluvial soils in climates where there is a distinction between dry and rainy seasons.
• The best temperature for growth is 80–97°F (27–36°C); there is no tolerance of frost and growth is poor above 2,300 feet (700 m).

Up to 50 ft. (15 m)

Up to 130 ft. (40 m)

Teak's ability to survive soaking, and a resistance to expansion or contraction between wetting and drying, made it a vital timber for ship construction, firstly in Southeast Asia and later for the European colonial powers. It is still used for ship's decking because it develops a nonslip surface—the weaker part of the annual growth rings wears first, allowing the narrower, stronger elements to stand proud.

Terminalia chebula

Yellow Myrobalan

A deciduous tree from India, Southeast Asia, and southern China that grows to a maximum height of 100 feet (30 m), the yellow myrobalan has small, hard, nutlike fruit that finds a number of uses in traditional cultures. The tree tends to be found in mixed deciduous forests in areas that experience a distinct dry season. It is well equipped to survive both fire and cutting because it regenerates readily from the base. Today, much of the fruit is still collected from wild trees, but there is also active management of natural woodland to maximize tree growth, and the tree is also grown in plantations. The fruit is much sought after by rodents and other wild animals, so commercial producers face strong competition. Various qualities of the fruit are recognized, based on the origin and degree of ripeness; those from Salem, in Karnataka, India, are thought the best.

Ayurvedic medicine makes use of the fruit for treating a wide variety of conditions. Little clinical research has been conducted on the fruit, but it does appear to contain antioxidant and antibacterial properties.

Myrobalan is very high in vitamin C. The tree yields nutlike fruits that are picked when green, pickled, and boiled in their own syrup with added sugar, or used in preserves. The fruit, which has a very distinctive scent and flavor, can also be made into a jam.

• Seed germination rates are poor, but seem to be improved by alternate wetting and drying in a nutrient solution.
• Root and shoot cuttings are also used.
• Young plants grow slowly, and thrive best if protected from full sun until they are several years old.
• The white flowers have a strong and unpleasant odor.

Up to 60 ft. (18 m)

Up to 100 ft. (30 m)

The main industrial use for the fruit, historically and today, is in tanning leather and the dyeing of fabrics. The fruit acts as a mordant in the dyeing process, helping to fix colors; it can also be used in its own right as a source of brown and yellow dyes. In silk production, extracts of the fruit are used to give body to the silk threads.

Terminalia ferdinandiana

Kakadu Plum

A small, semi-deciduous tree that reaches little more than 35 feet (10 m), the Kakadu plum is a northern Australian species. It produces the small, yellow-green fruit that currently holds the world record for vitamin C content— 100 times more than that of the orange. This is a good example of a plant that once was used only by indigenous peoples but which now has become fashionable with adventurous eaters and health-food aficionados. Most of the fruit is still wild-collected, often by Native Australian gatherers, possibly with a deleterious impact on the plant. Commercial cultivation, still in its early stages, suffers from the fact that plants grown in fertile conditions produce fruit with a lower vitamin content than those from more challenged wild plants.

Preparations made from the fruit were used by Native Australian people primarily to treat skin disorders or wounds. Clinical trials indicate that they have antibacterial properties.

Not only do Kakadu plums contain a lot of vitamin C, but also large quantities of antioxidants and minerals. They are too astringent to eat raw, but can be processed into jams, sauces, and relishes. When prepared in these ways, their flavor is said to resemble that of stewed fruit combined with citrus.

Up to 35 ft. (10 m)

Kakadu plums are now being used in cosmetic preparations, although there is little evidence to suggest that they offer any particular benefits for people who have normal and healthy skin.

Up to 35 ft. (10 m)

- The seed needs soaking in hot water for a few hours at least, to ensure removal of the remains of the fruit.
- The outer layer of the seed-coat then needs to be partly filed away to allow the seed inside to start to absorb water.
- Germination may then take many months.
- Once established, young plants can be expected to be very tolerant of drought, but not of frost or wet soils.

Theobroma cacao

Cacao, Cocoa Tree

The small evergreen tree that is distinguished as the source of one of the world's most cherished food products, cacao comes originally from the tropical regions of Central America and northern South America. It was cultivated and its fruit used by the great pre-Hispanic civilizations of Central America, and in the sixteenth century the Spanish invaders took it up and distributed it to the rest of the world. The largest proportion of raw cocoa is now produced in West Africa.

The tree grows to a height of only around 50 feet (15 m), and displays a habit very typical of tropical trees, that of bearing flowers and then fruits directly from the branches or the trunk. The fruit consists of a pod, up to 1 foot (30 cm) long, whose seeds are the source of the world's chocolate.

Traditional drinks in Central America and Mexico often contain cocoa, but are very different to the very sweet drink known to Europeans or North Americans as cocoa. Mexican *champurrado* is warm and thick, and is made with treated corn dough.

Cacao products contain many compounds that impact on the human body, most notably theobromine, which is related to caffeine. Its effects are subtle but undeniable; the Aztecs were known to smoke it in conjunction with tobacco.

Consumed as a bitter drink by the pre-Hispanic people of Central America, to whom it had great spiritual significance, cocoa was first made into what we know as solid chocolate during the early nineteenth century, in the Netherlands.

- Cacao seed has a very short period of viability, but germination and growth are both rapid.
- The plants need humid tropical conditions and can only be grown 10 degrees either side of the Equator.
- Shade is required, and cacao plantations are often integrated with existing forest trees.
- Cacao is an ideal crop for smallholder farming systems.

Up to 40 ft. (12 m)

Up to 50 ft. (15 m)

Thuja plicata

Western Red Cedar

This evergreen conifer grows to 230 feet (70 m) and forms a key part of the forested landscape of the Pacific Northwest of North America. It is a very important species in the timber trade. What sets the timber apart is its durability and its characteristic cedar smell, which indicates the presence of the compound thujaplicin. This compound acts as a fungicide and continues to perform that function for up to a century after the tree is felled. Thujaplicin almost certainly evolved as an adaptation to protect the tree in a very wet climate, and where the wood would otherwise rot rapidly. Unlike the dominant tree of the region, the Douglas fir (see p. 164), this species can regenerate in relatively deep shade.

The leaves, small and scalelike and densely clothing the branches, make the tree look to nonspecialists very much like one of the related cypresses. The cones are also unusually small, being less than ¾ inch (2 cm) wide. The tree's dense foliage has led to it being used extensively for hedging, for which it is very suitable because the growth rate is fast enough to give quick results, but not so fast that it easily gets out of control, like the notorious Leyland cypress, *Cupressus × leylandii*. It is also used as a forestry tree in climates similar to that of its homeland.

- Seed should germinate quickly and does not require stratification.
- Seedlings tend to be small and delicate at first, and benefit from light shade and constant moisture until they are established.
- The species shows great tolerance of a wide variety of soil types, including very shallow, infertile, and poorly drained soils.
- Being a wet-climate tree, it does not cope at all well with drought.

The timber's durability, along with the fact that it is relatively knot-free, means that it can be split into thin strips that are useful as cladding and roof shingles. Its ability to survive decay makes it ideal for posts, decking, and the construction of outdoor buildings, including beehives, where the avoidance of possibly toxic chemicals is vital. Although the timber is light (which is an advantage in making it a good insulator of sound and heat), it is remarkably strong—a good combination for construction purposes.

Native peoples of the Pacific Northwest used western red cedar leaf infusions for the treatment of colds, coughs, rheumatism, and other complaints. The tree is rarely used in modern herbalism, however.

Up to 105 ft. (32 m)

Up to 230 ft. (70 m)

The inner bark is edible and nutritious. Native American peoples would harvest and dry it, and then grind it into powder for use when traveling, or as an emergency food when there was nothing more palatable.

The soft, stringy bark has remarkable tensile strength, which made it an important raw material for Native Americans, who processed it into fibers for making rope, string, and yarn for weaving textiles. Clothes and blankets were made from the yarn derived from western red cedar bark.

. . . some tribes called themselves "People of the Cedar." Groves of ancient cedars were symbols of power, and gathering places for ceremonies, retreat, and contemplation.

JERI CHASE, "WESTERN REDCEDAR, 'TREE OF LIFE,'" ON OREGON DEPARTMENT OF FORESTRY WEBSITE

Basswood was a major source of fiber for prehistoric peoples and many tribes of American Indians.

"BASSWOOD," ON FOREST PRESERVE OF COOK COUNTY (IL) WEBSITE, NATURE BULLETIN NO. 422-A, JUNE 5, 1971

Tilia americana

Basswood, American Linden

Basswood is a deciduous tree that will grow to around 115 feet (35 m), forming a stately, domed shape. It is also known as American linden and American lime. In many respects, this species, which is found across eastern North America, is very similar to Eurasian species such as *Tilia cordata*. They all flourish on deep, fertile, and moist, or averagely moist, soils. All are climax species, but they tend to occur in relatively low concentrations, never forming pure stands. The name comes from "bast," the word given to the inner bark, which has been widely used for rope making by Native Americans.

Linden leaves are instantly recognizable because of their heartlike shape. The flowers are suspended from unusually shaped light-colored bracts. During early to mid-summer, the smell of their fragrant nectar can waft over considerable distances, signaling the trees' presence to wildlife. The flowers are an important food source for a variety of insect species.

While for centuries lindens have been popular street and parkland trees in Europe, in which context they are often pollarded, basswood is rarely used for such purposes. However, a number of selections, mostly according to shape, have been made for landscape use.

- *Tilia* species generally are not easy to propagate from seed, as viability is very low, and unless sown immediately, deep dormancy sets in and germination may take up to two years.
- In the wild, regeneration often occurs through suckering.
- Cuttings are the most usual method of propagation.
- The tree tolerates a little shade, needs moist soils, and can be heavily pruned.
- Susceptible to adult Japanese beetles, which eat its leaves.

All species of the linden have been exploited for bast, the tough fibrous layer just below the bark. Native Americans (and many other preindustrial peoples) used comparable techniques: sheets of material were soaked in water for several weeks until the connecting tissue rotted, leaving the very strong fibers. These could then be fashioned into rope, fishing lines, and thread. This was once a dominant use of the Eurasian species, too, but the skills needed to work the bast have died out.

The pale brown wood, sometimes white or tinged with red, is among the lightest and softest of all hardwoods. The commercial uses of basswood are thus largely limited to veneers, plywood, and pulping.

Young linden leaves are pleasant to eat, either raw or lightly cooked, soon after they emerge in spring. The bast can also be eaten fresh, or dried and ground into flour for adding as a thickener in stews and soups. Linden tea, made from the flowers, has a pleasing flavor.

Basswood, like all lindens, has a dense and even grain. It takes cuts well in all directions, which makes it ideal for carving. Europe's lindens provided the raw material for some of the best woodcarving in its history.

Up to 80 ft. (25 m)

Up to 115 ft. (35 m)

Ulmus procera

English Elm, European Elm

The so-called English elm offers an extraordinary (and salutary) story. Once an important part of the English cultural landscape, and to a lesser extent, other parts of northern Europe too, the tree was hit very badly by Dutch elm disease in the late 1960s. Huge areas were completely transformed as the magnificent trees—often around 100 to 130 feet (30–40 m)high—began to die. Evidence from fossil pollen suggested that there have been previous die-offs since the last ice age. Today, the species survives, as a hedgerow tree, but once it gets to around 15 feet (5 m), it is discovered by the beetles that spread the fungus, which will then knock the tree back to ground level again. Dutch elm disease is also spread through the roots; in hedgerows, roots from individual trees often link up through root grafts.

In 2004, research was published that showed that all the examples of this species in Britain were actually one single clone, which all shared the same vulnerability to disease. It was proposed that the Romans brought it from Spain to England to provide poles on which to train vines. The Roman agronomist Columella, in a work on farming, suggests using elm, in particular one variety of Italian origin, which did not set seed, and was therefore sterile—and this is possibly the ancestral variety.

- Seed is sterile, but the tree is easy to propagate—cut a large twig, place it in the ground, and up it goes.
- Many country hedges were planted this way, and there is no reason why the species should not continue to be used for hedging, especially for deep fertile soils.
- A number of supposedly disease-resistant cultivars are available—none of them are likely to be 100 percent disease resistant.
- Natural regeneration is entirely by root suckers.

The inner bark was traditionally used by European country herbalists to treat digestive problems and wounds. It is distinctly mucilaginous (viscous and sticky) and has anti-inflammatory properties. The leaves have a pleasant flavor and have been used as a mouthwash to freshen the breath.

The fibers from the inner bark of the European elm are extremely tough and were once widely used for rope making—with these often being extracted from the branches rather than the trunk of the tree. The American elm (*Ulmus americana*) was used in a very similar way by Native American peoples.

Up to 80 ft. (25 m)

Up to 115 ft. (40 m)

Elm wood is strong and remarkably resistant to decay when wet (but, oddly, not when buried). It is also resilient to crushing. It was popular in preindustrial times for hollowing out and making water pipes and was also used in the construction of jetties and piers. It warps and shrinks badly on drying, however, and consequently was never used for major structural work. It was also used in providing some components for shipbuilding, farm buildings, and in making waterwheels. Sometimes it was used for making furniture but not as much as oak.

> *The bright golden color of the lines of elms in the hedgerows is one of the most striking scenes that England can produce.*
>
> HENRY ELWES, *THE TREES OF GREAT BRITAIN AND IRELAND, VOL. VII* (1913)

Ulmus rubra

Slippery Elm, Red Elm

Very similar to the American elm (*Ulmus americana*), and at times considered simply a variety of it, this tree has a similar, eastern North American, distribution. Growing to 6 feet (20 m), it is distinguished by its hairier leaves and more upright branching pattern. Both the slippery elm and its close relative are trees of fertile, moist situations, particularly river flood plains. Both are common, generally mixed with maple species and ash, never growing in pure stands.

Whereas American elm was very popular in the landscape, and in particular as a street tree, in the early twentieth century, this species has always been seen as somewhat unattractive, even "weedy." It has never been developed commercially and, like its cousin, it suffers from Dutch elm disease.

Slippery elm bark is one of the most versatile of North American herbal treatments. The mucilaginous inner bark can be boiled with water to produce a healing gel for wounds, burns, and skin ailments. The leaves may be made into a tea as a treatment for digestive problems. Bark jelly is recommended to those who are unable to eat solid food.

The wood is hard, but historically it has tended to have been seen as being of low quality. Typical uses were for farm implements and wagon wheels, especially their hubs. It was also used as an alternative to yew for bows.

Up to 60 ft. (18 m)

Up to 65 ft. (20 m)

> *Slippery elm bark possesses also great influence upon diseases of the female organs.*
>
> M. GRIEVE, "SLIPPERY ELM,"
> ON THE BOTANICAL WEBSITE

- Seed should be sown immediately it is ripe, when germination is usually rapid.
- Stored seed requires several months of stratification.
- Young plants develop a taproot; they should be moved into deep pots or into a nursery bed, and planted out within two years.
- Suckers from existing rootstocks can also be detached and planted.
- Young trees are unaffected by Dutch elm disease.

Umbellularia californica

Californian Laurel, Headache Tree

This evergreen species of coastal California makes an impressive tree, reaching up to 100 feet (30 m). Mostly, it grows alongside California redwood and pine species, and sometimes oak. It is regarded as a remnant of the extensive evergreen "laurel forest" that would have covered southwest North America during the Tertiary era. The tree may vary considerably in size and shape depending on the nature of its environment.

The most distinctive aspect of this species is the intense aroma from oils contained in the long, narrow leaves; breathe in too deeply and a headache can result. Insignificant flowers mature to form fruits 6½ feet (2 m) long that look like miniature avocados (a close relative). The fruits contain a single seed.

The fruit has been used as a food by Native Americans. More attractive are the nutlike seeds, which can be cracked, roasted, and eaten. They have also been ground and used as a coffee substitute.

The wood is fine-grained and hard, and is good for turning and making into small objects. Historically, it has been used for furniture. It has good tone and is sometimes used for acoustic musical instruments.

Up to 30 m (100 ft.)

Up to 30 m (100 ft.)

A relative of the familiar bay laurel, it has leaves that may be used for seasoning in cooking. Cooks must allow for the fact that they are very much stronger and spicier than bay, with a pungent, camphorlike odor.

- Seed should ideally be sown fresh. Germination may take many months.
- The slow-growing seedlings have taproots and are best container-grown until large enough to plant out.
- The tree flourishes in most well-drained soils in sun, or light shade, and is reasonably drought-tolerant.

Japanese Pepper

Growing to a maximum height of only 15 feet (5 m), the Japanese pepper is a small deciduous tree or large shrub from Japan, China, and Korea. It is a member of a large genus containing around 250 species, and very similar trees may be found in North America, say, such as *Zanthoxylum americanum* (common pricklyash). Because the various *Zanthoxylum* species tend to be very similar, misidentification is common.

In the Japanese pepper, small, pinnate leaves grow from spiny twigs, although forms have been selected that are thornless, to make for easier harvesting of the fruits. These are ¼ inch (5 mm) wide and have a peppery flavor, with a slightly numbing effect on the mouth, which is made great use of in Far East Asian cuisines. The flavor comes from small oil glands, which are typical of the Rutaceae family, and these can be seen if a leaf is held up to the light.

- Seed requires fall sowing, or three months of stratification
- The seed is hard, and scarifying with sandpaper may promote germination.
- Stem and root cuttings may be used for propagation.
- The tree tolerates some shade, and thrives in any reasonably fertile soil.
- The tree may sucker in time.

The fruits' sharp and rather surprising flavor is used to great effect in enhancing certain dishes, such as eel. While it is the dried fruit that is the commercially important product, the fresh leaves are often sold as a vegetable in spring wherever it grows, especially in Japan.

The fruits are used to treat a variety of ailments in traditional medicine, as are those of the American species. The numbing quality is effective against toothache, hence *Z. americanum*'s common name of toothache tree.

The narrow but strong trunks are used to make walking sticks in East Asia, as well as implements that need to be hard and resilient, such as traditional pestles.

Up to 15 ft. (4.5 m)

Up to 15 ft. (4.5 m)

Jujube

This fast-growing deciduous tree attains 40 feet (12 m) and comes from arid regions of central Asia, although a long time in cultivation has made its original distribution unclear. Having been grown in parts of China for 4,000 years, the tree spread along historical trade routes, and in some places, such as Madagascar, it has become a seriously invasive species. The fruits are up to 1 inch (3 cm) across, with a central pitlike seed. A closely related species, *Ziziphus mauritiana* has inferior fruit which nevertheless has a very high vitamin C content. Both species show remarkable environmental tolerance, surviving extremes of heat and cold, which is probably a major reason for *Z. jujuba*'s popularity in much of Asia, and its long history of cultivation in China. Eaten fresh, the fruit tastes of apple; it wrinkles as it dries, and in this form it keeps almost indefinitely, a boon to nomads. Traditional varieties, mostly of Chinese origin, are now being joined by ones bred in the United States, where they thrive in drier areas.

Jujubes are used in several traditional Asian medical systems, both for treating physical ailments and for relieving anxiety and insomnia—the seeds contain flavonoids with sedative properties. A tea made from the leaves is used to soothe sore throats.

Like dates, jujube fruits are mostly eaten dried, rather than fresh, very often as a simple snack. There are many different traditional ways of preserving them, including pickling or salting them on their own, or pitting them and the mixing the resulting flesh with spices.

- Seed needs to be sown in fall, or given at least three months of stratification after a warming period of three months.
- Propagation from cuttings and suckers is possible.
- The tree needs well-drained soils, although it tolerates infertility and drought.
- Plants are hardy to -4°F (-20°C), but only in drier areas, and long, hot summers are necessary for fruit production.

Up to 25 ft. (8 m)

Up to 40 ft. (12 m)

The wood is hard and strong, but does not appear in quantity in the timber trade. It is often used for carving or turning, in the production of small objects such as bowls, or pieces for board games such as chess, checkers, mahjong, and Go.

Glossary

Acorn The nut of an oak species, *Quercus*, inevitably held inside a cup, which is attached to a twig.

Alkaloid Nitrogen-containing compounds made by plants (but also by other living things too), which nearly always have a bitter taste, and sometimes a potent reaction, and are sometimes toxic. Many of the active ingredients in herbal medical treatments are alkaloids.

Allelopathic Certain plants produce chemical compounds that inhibit the growth of other plants around them— a potent way of reducing local competition.

Ayurvedic "Ayurveda" is a traditional system of medicine and health management developed in India over many thousands of years. Its use of herbal products is very sophisticated, but it is not necessarily evidence-based.

Calcareous This refers to soil conditions that are alkaline, or basic, i.e., the opposite of acidic. Soil alkalinity and acidity has an impact on the availability of nutrients in the soil, which may have a strong effect on the plant species that may grow there.

Canopy This refers to the mass of the trees in a forest, where most of the tree foliage is within a certain height range. A minority of emergents may stand above it, and more sparse layer of understory species below it.

Carcinogenic Cancer-inducing. Some chemicals produced by plants may, in susceptible individuals, if ingested over long periods, promote cell-damage followed by the development of cancers.

Climax The final stage of succession, where a mature forest community achieves relative stability. See Introduction.

Coppice The regular cutting down to ground level of tree species which recover quickly from such cutting. The long straight vigorous shoots that the stump sends up are often very valuable material for traditional crafts.

Cultivar A variety of a plant where the individuals are genetically identical,

and that has been selected for particular desired characteristics, such as large and tasty fruit. Cultivars are usually propagated by grafting or cuttings.

Dominant A tree species that dominates the mature, climax, forest community.

Emergent See canopy.

Genus The first part of the name in the Linnaean system of Binomial Naming (often called scientific or Latin names). A genus gathers together obviously closely-related species, and may comprise between one or hundreds of species.

Grafting The practice of attaching a cutting of one variety (or cultivar) onto a closely-related variety with an established root system, so propagating the variety and joining together in one plant two different and distinct genomes. The rootstock (i.e., the one with the roots) will control the rate of growth of the attached variety.

Inflorescence A flower head. Popular parlance often refers to such a head of flowers or compound flowers as if they were individual flowers.

Invasive Having the tendency to spread vigorously, and suppressing the growth of other plant species. Usually applied to alien, i.e., non-native species, but native species can be invasive in some circumstances, in their own natural environment.

Mature Either a tree that has reached something approaching its maximum size and sexual maturity, i.e., flowering and fruiting, or a forest community that has reached the climax stage.

Naturalize The tendency of some plants species to become wild or feral, in regions where they are not native, i.e., not found naturally. This is not necessarily problematic, and only become so when species become invasive.

Nitrogen-fixing Having the ability to turn atmospheric nitrogen molecules into water-soluble nitrogen-containing

compounds, which plants can then access as a nutrient. Plants cannot do this, but some have evolved to form symbiotic relationships with bacteria in their roots, which can perform this quite challenging chemical trick.

Palmate With leaves that are approximately hand-shaped, with lobes radiating out from where the body of the leaf joins the stem.

Pinnate Where leaves are divided into at least three leaflets, each leaflet being a discrete part of the body of the leaf connected to the other leaflets by a central stem. Leaflets are nearly always opposite each other on this stem.

Pioneer The first plant species to occupy bare ground. With trees, it refers to the first tree species to establish in either bare ground or an existing, but non-woody, plant community. Pioneers are often short-lived, but not always; some may go on to be part of the climax community.

Pith The soft center of the twigs of some tree species, generally those that aim to maximize volume of growth over weight, such as fast-growing pioneer species.

Pollard See coppice. The tree is kept cut to at least head height, often higher.

Rootstock See grafting.

Scarify/scarification See section on Cultivation.

Subspecies A division of a species, where relatively small differences define a particular geographically defined population as distinct.

Suckers The shoots thrown up by tree roots, each one of which is capable of developing into an independent tree if the main trunk is damaged. Suckers may appear some way from the base of the tree, but are often only immediately around the base.

Tincture A solution of active ingredients in alcohol.

Understory See canopy.

Index

Picture Credits & Acknowledgments

Quintessence would like to thank Cheryl Hunston for the index, Jodie Gaudet for americanizing, and Lesley Malkin for proofreading. Every effort has been made to credit the copyright holders of the images used in this book. We apologize for any unintentional omissions or errors and would be pleased to insert the appropriate acknowledgment to any companies or individuals in any subsequent editions of the work.

Alamy, De Agostini Picture Library / Contributor 37b anawat sudchanham 43 Bob Bampton 24r, 152r De Agostini Picture Library 24l, 46l, 52r, 55l, 61l, 88, 95, 97l, 100, 103, 108l, 112b, 131 De Agostini Picture Library / Contributor 25l, 27, 40, 44, 45, 57r, 59, 140, 150l, 157, 159, 173l, 182b, 194l, 206l, 213 Denys Kurylow 72l, 98r, 136 © DK Limited/CORBIS 53r, 117 © Dorling Kindersley 57l, 77l, 79, 84r, 107l, 127, 155, 163l, 214, 217l, 218b EMJAY SMITH 80 gyn9037 66l Jakkrit Orrasri 39, 122 Jan Martin Will 171 Kati Aitken 49, 113, 119, 124, 144, 178-9, 199, 208 © Lano Angelo | Dreamstime.com 20l Loupe Project 114l Maksym Bondarchuk 118l Malgorzata Kistryn 85b © markku murto/art / Alamy 195r Mary Evans / Natural History Museum 38 Michelle Ross 16, 19 niceregionpics 31 polaris50d 29l Potapov Alexander 121 Sue Oldfield 12l, 62r The Art Archive / DeA Picture Library 25r, 56, 151r, 154 The Art Archive / DeA Picture Library / A. De Gregorio 170 The Art Archive / DeA Picture Library / G. Cigolini 82r The Art Archive / Kharbine-Tapabor / Collection LOU 37t The Art Archive / Kharbine-Tapabor / Florilegius 162t © The Natural History Museum 36 Tim Hayward 28l, 19 ukmooney 175 UniversalImagesGroup / Contributor 163r Zelimir Borzan, University of Zagreb, Bugwood.org 61r, 215 Zerbor, Maksym Bondarchuk 21r ZQPhotography 160

Plants for a Future (www.pfaf.org) has been a very useful first port of call in researching plant uses. It does valuable work and deserves our support. Sources are always given, so it is a good way to check the accuracy of information.